Common Core Math Activities

AUTHORS: KARISE MACE and CHRISTINE HENDERSON
EDITOR: MARY DIETERICH
PROOFREADER: MARGARET BROWN

COPYRIGHT © 2015 Mark Twain Media, Inc.

ISBN 978-1-62223-532-2

Printing No. CD-404235

Mark Twain Media, Inc., Publishers
Distributed by Carson-Dellosa Publishing LLC

Table of Contents

Introduction to the Teacher

People are naturally curious. When educators can engage their students' curiosity and guide them through meaningful activities, valuable learning takes place. The math labs included in this book are designed to do just that!

These lab activities are hands-on and allow students to explore and gain deeper understanding of mathematical concepts. From packing and wrapping packages to crime scene investigation, students will be challenged to pull from prior mathematical knowledge and extend it as they investigate mathematical relationships and concepts. In addition to being fun, these creative activities will help students become better problem solvers.

Each of the lab activities is aligned to multiple Common Core State Standards for Mathematics. This correlation is included on the teacher page at the beginning of each lab. The teacher page also includes a materials list, a pacing guide, and helpful tips for how to make the lab a successful learning experience.

We hope that you and your students enjoy exploring the world of mathematics as you work on these activities together!

—Karise Mace and Christine Henderson

Correlation of Labs to Common Core State Standards for Mathematics

Grade 6 Standards and Labs	Lab 1	Lab 2	Lab 3	Lab 4
6.G.A.1			•	
6.G.A.2	•			
6.G.A.3			•	
6.G.A.4	•			
6.RP.A.1	•			•
6.RP.A.2				•
6.RP.A.3	•			•
6.NS.A.1	•			
6.NS.B.2				•
6.NS.B.3				•
6.NS.B.4				
6.NS.C.5		•		
6.NS.C.6		•		
6.NS.C.7		•		
6.NS.C.8				
6.EE.A.1				
6.EE.A.2			•	
6.EE.A.3				
6.EE.A.4				
6.EE.B.5			•	
6.EE.B.6			•	
6.EE.B.7			•	
6.EE.B.8				
6.EE.C.9	•			
6.SP.A.1				
6.SP.A.2		•		
6.SP.A.3		•		
6.SP.B.4		•		
6.SP.B.5		•		

Grade 7 Standards and Labs	Lab 1	Lab 2	Lab 3	Lab 4
7.G.A.1				•
7.G.A.2				•
7.G.A.3				
7.G.B.4	•			
7.G.B.5				
7.G.B.6	•			
7.RP.A.1				
7.RP.A.2		•	•	
7.RP.A.3			•	
7.NS.A.1		•		
7.NS.A.2	•	•		•
7.NS.A.3	•	•		•
7.EE.A.1				
7.EE.A.2			•	
7.EE.B.3			•	
7.EE.B.4			•	
7.SP.A.1		•		
7.SP.A.2		•		
7.SP.B.3		•		
7.SP.B.4		•		
7.SP.C.5	•			
7.SP.C.6	•			
7.SP.C.7	•			
7.SP.C.8	•			

Grade 8 Standards and Labs	Lab 1	Lab 2	Lab 3	Lab 4
8.G.A.1				
8.G.A.2		•		
8.G.A.3		•		
8.G.A.4		•		
8.G.A.5				
8.G.B.6				
8.G.B.7		•		
8.G.B.8		•		
8.G.C.9				•
8.NS.A.1		•		•
8.NS.A.2		•		•
8.EE.A.1			•	
8.EE.A.2		•		
8.EE.A.3			•	
8.EE.A.4			•	
8.EE.B.5	•			
8.EE.B.6				
8.EE.C.7				
8.EE.C.8	•			
8.F.A.1	•			
8.F.A.2	•			
8.F.A.3	•			
8.F.B.4	•			
8.F.B.5				
8.SP.A.1	•			
8.SP.A.2	•			
8.SP.A.3	•			

Grade 6, Lab #1: Packing and Wrapping

Teacher Information

Introduction

In this lab, students will explore volume and surface area of rectangular prisms and how changing the dimensions of the prisms affects the surface area and volume.

Duration of Lab

3 days

Common Core State Standards

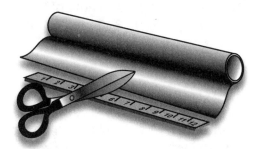

- 6.G.A.2
- 6.G.A.4
- 6.RP.A.1
- 6.RP.A.3
- 6.NS.A.1
- 6.EE.C.9

Prerequisite Skills

Before completing this activity, students need to be able to:
- calculate the volume of a rectangular prism.
- calculate the area of a rectangle.
- measure lengths with a centimeter ruler.

Supplies

Each STUDENT will need...	Each GROUP will need...	The TEACHER will need...
• Scissors • A centimeter ruler • A copy of the student pages	• Scotch tape • Masking tape • Copies of nets on cardstock	• Wrapping paper • Scissors • Small candies • One class data sheet

Teacher Tips

- This lab works best in pairs or small groups.
- Make the necessary copies of the package nets for each group. Copy the nets onto cardstock heavy enough to hold its shape when folded.
- What students choose to label as length, width, and height may vary. As long as the three measurements for each box are correct, they are acceptable.
- The column headings have not been provided for students in the tables for Days 1 and 2. However, sample answers are provided in the key.
- When students ask for a piece of wrapping paper on Day 2, make sure to cut the piece to the exact dimensions that they specify.
- You may wish to reward the group(s) who does the most efficient and neatest package wrapping job on Day 2.

Name: _____ Date: _____

Grade 6, Lab #1: Packing and Wrapping

In this lab, you will work with at least one partner to pack and wrap some packages. You have three main goals:

1. Determine which package will hold the most.
2. Wrap one package using the least amount of wrapping paper possible.
3. Discover the relationships between dimensions, surface area, and volume of rectangular prisms.

Day 1: Fill it Up

Part 1: Build the Packages

Cut out each of the nets. Fold and tape to form rectangular prisms.
***Note: Do NOT tape the packages completely closed. See notes on nets.

Part 2: Determine Which Package Will Hold the Most

1. Without measuring, predict which package will hold the most. Write your hypothesis and at least one sentence explaining your choice.

   _____

   _____

2. How can you test your hypothesis? _____

   _____

3. Calculate which package will hold the most. Organize your data and work in the table below.

Package	Length (cm)	Width (cm)	Height (cm)	Volume (cm³)
A				
B				
C				
D				

4. Was your hypothesis correct? Why or why not? _____

   _____

Part 3: Check Your Results

1. Take the package that your group calculated will hold the most to your teacher. Your teacher will fill this package with candies.
2. Return to your desk. Count and record the number of candies in your package.

   Number of candies: _____

3. Record the data about your filled package on the class data sheet.
4. On your own paper, write a short paragraph summarizing the results of your experiment and how your results compare with your classmates.

Name: _____ Date: _____

Grade 6, Lab #1: Packing and Wrapping

Day 2: All Wrapped Up

Part 1: Determine Which Package Can Be Wrapped With the Least Amount of Paper

1. Predict which package can be wrapped using the least amount of wrapping paper. Write your hypothesis and at least one sentence explaining your choice.

   _____

   _____

   _____

2. How can you test your hypothesis? _____

   _____

   _____

3. Copy your measurements from Day 1 into the table below. Then calculate the amount of wrapping paper necessary to wrap each package and record those values in the table.

Package	Length (cm)	Width (cm)	Height (cm)	Surface Area (cm^2)
A				
B				
C				
D				

4. Was your hypothesis correct? Why or why not? _____

   _____

   _____

5. Select the package that would require the smallest amount of wrapping paper to cover it. Then determine the dimensions of one rectangular shaped piece of wrapping paper that you could use to wrap this package with the least amount of overlap or waste.

Part 2: Check Your Results
1. Ask your teacher for a piece of wrapping paper with the dimensions you determined in Part 1.
2. Return to your desk. Wrap your package with the piece of wrapping paper provided.
3. Place your wrapped package in the display space indicated by your teacher.
4. Record the data about your wrapped package on the class data sheet.
5. On your own paper, write a short paragraph summarizing the results of your experiment and how your results compare with your classmates.

Name: _____ Date: _____

Grade 6, Lab #1: Packing and Wrapping

Day 3: Short and Wide or Tall and Narrow?

Analyzing the Data

1. Complete the following table.

Package	Length (cm)	Width (cm)	Height (cm)	Surface Area (cm²)	Volume (cm³)
A					
B					
C					
D					

2. What is the relationship between the dimensions of

    a. Packages A and B? _____

    _____

    b. Packages C and D? _____

    _____

3. Compare the surface areas of the packages listed below. Express the relationship between their surface areas as a simplified ratio.

    a. Packages A and B _____

    _____

    b. Packages C and D _____

    _____

4. Explain the relationship between the dimensions of rectangular prisms and their surface areas.

    _____

    _____

    _____

    _____

5. Write an equation to express the relationship between the surface area of one rectangular prism and another rectangular prism whose dimensions are twice that of the first prism. Let x represent the surface area of the smaller prism and y represent the surface area of the larger prism.

    _____

Grade 6, Lab #1: Packing and Wrapping

Day 3: Short and Wide or Tall and Narrow?

Analyzing the Data (cont.)

6. Compare the volumes of the packages listed below. Express the relationship between their volumes as a simplified ratio.

    a. Packages A and B _____

    _____

    b. Packages C and D _____

    _____

7. Explain the relationship between the dimensions of rectangular prisms and their volumes.

    _____

    _____

    _____

8. Write an equation to express the relationship between the volume of one rectangular prism and another rectangular prism whose dimensions are twice that of the first prism. Let *a* represent the volume of the smaller prism and *b* represent the volume of the larger prism.

    _____

9. Suppose that Package E has dimensions that are twice those of Package F.

    a. If the surface area of Package F is 64 square inches, what is the surface area of Package E? Explain how you determined your answer.

    _____

    _____

    _____

    b. If the volume of Package E is 576 cubic inches, what is the volume of Package F? Explain how you determined your answer.

    _____

    _____

    _____

Name: _____ Date: _____

Grade 6, Lab #1: Packing and Wrapping

Class Data Collection Sheet

Day 1: Fill it Up

Group	Package	Volume	Number of Candies

Day 2: All Wrapped Up

Group	Package	Surface Area	Dimensions of Wrapping Paper

Grade 6, Lab #1: Packing and Wrapping

Net for Package A

Remember to
leave one side
open so you
can fill the box.

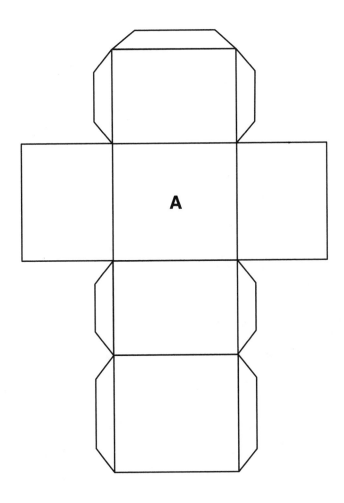

Grade 6, Lab #1:
Packing and
Wrapping

Net for Package B

B

Remember to
leave one side
open so you
can fill the box.

Grade 6, Lab #1: Packing and Wrapping

Net for Package C

Remember to
leave one side
open so you
can fill the box.

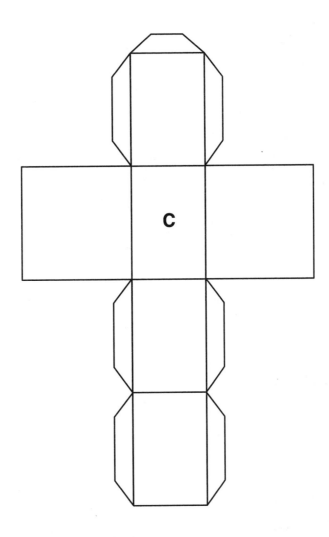

Grade 6, Lab #1: Packing and Wrapping

Net for Package D

Remember to
leave one side
open so you
can fill the box.

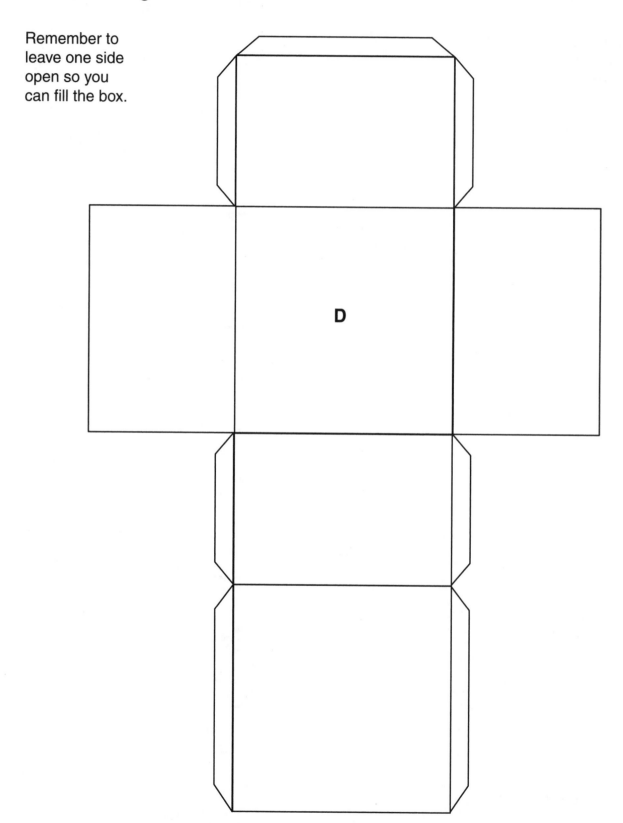

Grade 6, Lab #2: It's Freezing Around Here!

Teacher Information

Introduction

In this lab, students will explore integers and the relationships between them in a real-life context. Further, students will create statistical data displays and analyze the distributions of temperature data.

Duration of Lab

3 days

Common Core State Standards

- 6.NS.C.5
- 6.NS.C.6
- 6.NS.C.7
- 6.SP.A.2
- 6.SP.A.3
- 6.SP.B.4
- 6.SP.B.5

Prerequisite Skills

Before completing this activity, students need to be able to:
- plot rational numbers on a number line.
- create a dot plot.
- determine mean, mean absolute deviation, median, and interquartile range.

Supplies

Each STUDENT will need...	Each GROUP will need...
• Copy of the daily lab sheets	• 31 small sticky notes in two or three different colors • Two or three long pieces of masking tape • A black or blue permanent marker • A computer with Internet access

Teacher Tips

- This lab works best in pairs or small groups.
- Before introducing Day 3, be sure to check the website provided (http://www.usclimatedata.com/) to make sure that it is still active. It's also a good idea to familiarize yourself with the organization of the website and how students can use it to access the data they need.
- You may wish to have your students make a sticky note dot plot of the local data as well and compare it to the other dot plots.
- If you do not want students to place masking tape on the wall or floor, you may provide poster board for them to use when creating their sticky note dot plots.

Name: _____   Date: _____

Grade 6, Lab #2: It's Freezing Around Here!

In this lab you will work with a partner or in a small group to analyze temperature data. You will:
1. Collect data.
2. Identify trends in data.
3. Calculate measures of center and variability.
4. Graph data.

Day 1: Getting Up Close and Personal

Part 1: Grise Fiord, Nunavut, Canada

Way up in the Arctic Circle there is a Canadian Island named Ellesmere Island. Grise Fiord is the largest community on Ellesmere Island and has a population of about 130 people.

The data in the table shows the mean temperature for Grise Fiord each day in May of 2014.

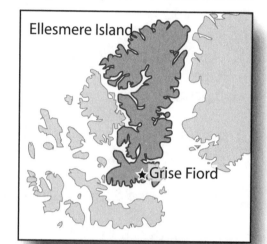

Day	Temperature (°C)
1	−10.8
2	−14.1
3	−16.2
4	−14.5
5	−12.8
6	−11.6
7	−10.5
8	−8.0
9	−6.0
10	−9.2
11	−11.5

Day	Temperature (°C)
12	−9.8
13	−10.9
14	−11.3
15	−7.1
16	−5.2
17	−8.8
18	−6.6
19	−7.5
20	−3.9
21	−5.1

Day	Temperature (°C)
22	−6.4
23	−6.5
24	−4.5
25	−2.7
26	−1.8
27	−4.8
28	−3.1
29	−5.6
30	−5.5
31	−9.2

1. Spend a few minutes just looking at the data in the table. Then write three observations you can make about the data. Compare your observations with those that your partner made.

_____

_____

_____

2. What was the highest temperature in Grise Fiord during May 2014? Explain how you determined your answer. _____

_____

_____

Name: _____ Date: _____

Grade 6, Lab #2: It's Freezing Around Here!

Day 1, Part 1 (cont.)

3. What was the lowest temperature in Grise Fiord during May 2014? Explain how you determined your answer. _____

_____

_____

4. Plot the highest and lowest temperatures on the number line. Then estimate the range of temperatures in Grise Fiord during May 2014. Explain how you determined your answer.

_____

_____

_____

Part 2: Toronto, Ontario, Canada

Just across Lake Ontario from New York State is Toronto. It is the most populated city in Canada with a population of approximately 2.6 million people.

The data in the table shows the mean temperature for Toronto each day in May of 2014.

Day	Temperature (°C)	Day	Temperature (°C)	Day	Temperature (°C)
1	10.8	12	17.3	22	17.0
2	9.3	13	16.0	23	13.0
3	10.3	14	16.3	24	16.5
4	9.3	15	16.3	25	17.8
5	9.5	16	9.8	26	22.8
6	10.8	17	8.0	27	24.3
7	9.8	18	9.0	28	15.3
8	14.5	19	13.3	29	16.3
9	18.0	20	15.0	30	17.8
10	17.5	21	16.8	31	16.3
11	14.0				

Name: _____ Date: _____

Grade 6, Lab #2: It's Freezing Around Here!

Day 1, Part 2 (cont.)

1. Spend a few minutes just looking at the data in the table. Then write three observations you can make about the data. Compare your observations with those that your partner made.

   _____

   _____

   _____

2. What was the highest temperature in Toronto during May 2014? Explain how you determined your answer. _____

   _____

3. What was the lowest temperature in Toronto during May 2014? Explain how you determined your answer. _____

   _____

4. Plot the highest and lowest temperatures on the number line. Then estimate the range of temperatures in Toronto during May 2014. Explain how you determined your answer.

   _____

   _____

   _____

Part 3: Comparing Temperatures

1. On which day in May were the temperatures in Grise Fiord and Toronto opposite of each other? Explain how you determined your answer.

   _____

   _____

2. Were there any other incidences of opposite temperatures between Grise Fiord and Toronto, not necessarily on the same day?

   _____

   _____

Name: _____ Date: _____

Grade 6, Lab #2: It's Freezing Around Here!

Day 1, Part 3 (cont.)

3. On average, was it colder in May in Grise Fiord or Toronto? Explain your reasoning.

_____

_____

_____

Day 2: A Summary of the Cold Spots

Part 1: Dot Plots of the Cold Spots

1. Write the temperature for each day in May 2014 for Grise Fiord on one color of sticky note. Each temperature should be on its own sticky note.

2. Write the temperature for each day in May 2014 for Toronto on another color of sticky note. Again, each temperature should be on its own sticky note.

3. Make two number lines using pieces of masking tape. You will use these to create dot plots of the temperature data for both cities. Make sure you use the same intervals on both number lines. Place one above the other, taking care to align the zeros, on the table, floor, or wall as directed by your teacher. Be sure to leave enough space between the lines to place the sticky notes. Create a dot plot of the May 2014 temperature data for Grise Fiord on the top number line and for Toronto on the bottom number line.

4. How do your number lines compare to those of your classmates?

_____

_____

_____

5. Write a couple of sentences explaining what the dot plots tell you about each of the following. Then discuss your answers with your class.

   a. the distribution of the temperatures in Grise Fiord during May 2014

   _____

   _____

   _____

   b. the distribution of the temperatures in Toronto during May 2014

   _____

   _____

   _____

   c. the comparison of temperatures between Grise Fiord and Toronto during May 2014

   _____

   _____

   _____

Name: _____ Date: _____

Grade 6, Lab #2: It's Freezing Around Here!

Day 2: A Summary of the Cold Spots

Part 2: The Cold, Hard Facts

1. Which measures of center and variability would best describe the temperature data for May 2014 in Grise Fiord? Explain your reasoning.

   _____

   _____

   _____

2. Calculate the measures of center and variability that you chose in Exercise 1. Then, compare your results with your classmates.

   _____

   _____

   _____

3. Write one sentence explaining what the measures of center and variability tell you about the temperature distribution for May 2014 in Grise Fiord.

   _____

   _____

   _____

4. Which measures of center and variability would best describe the temperature data for May 2014 in Toronto? Explain your reasoning.

   _____

   _____

   _____

5. Calculate the measures of center and variability that you chose in Exercise 4. Then compare your results with your classmates.

   _____

   _____

   _____

6. Write one sentence explaining what the measures of center and variability tell you about the temperature distribution for May 2014 in Toronto.

   _____

   _____

   _____

Name: _____   Date: _____

Grade 6, Lab #2: It's Freezing Around Here!

Day 3: How's the Weather At Home?

Part 1: Collecting Local Temperature Data

1. Go to http://www.usclimatedata.com/ and download the daily temperature data for May 2014 for your city or the city closest to you. Record it in the table below. (HINT: Be sure to record the temperatures in degrees Celsius.)

Day	Temperature (°C)	Day	Temperature (°C)	Day	Temperature (°C)
1		12		22	
2		13		23	
3		14		24	
4		15		25	
5		16		26	
6		17		27	
7		18		28	
8		19		29	
9		20		30	
10		21		31	
11					

2. Make a dot plot of your temperature data.

3. Which measures of center and variability would best describe the temperature data for May 2014 in your city? Explain your reasoning.

    _____

    _____

4. Calculate the measures of center and variability that you chose in Exercise 3. Then compare your results with your classmates.

    _____

    _____

Part 2: Shall I Spend Next May at Home or in Canada?

1. On your own paper, write a short paragraph about whether you would like to spend next May in Grise Fiord, Toronto, or your city. Use the data you analyzed in this lab to support your answer.

Grade 6, Lab #3: Crazy Quilt

Teacher Information

Introduction
In this lab, students will explore area of composite figures. They will use equations for calculating areas and will graph geometric figures in the coordinate plane.

Duration of Lab
3 days

Common Core State Standards
- 6.G.A.1
- 6.G.A.3
- 6.EE.A.2
- 6.EE.B.5
- 6.EE.B.6
- 6.EE.B.7

Prerequisite Skills
Before completing this activity, students need to be able to:
- plot points in a coordinate plane.
- determine the length of a line in the coordinate plane whose endpoints have the same first coordinate or the same second coordinate.
- use area formulas for triangles and quadrilaterals.

Supplies

Each STUDENT will need...	Each GROUP will need...	The TEACHER will need...
• Copy of the daily lab sheets • Copy of grid paper • Straightedge	• Copy of Crazy Quilt Square • Pair of scissors • Colored pencils • A variety of material • One 18 cm by 18 cm square of cardstock • Glue • Straight pins	• Books with pictures of quilts (optional) • Scissors • Centimeter ruler

Teacher Tips
- This lab works best in pairs.
- If you are unable to obtain a variety of fabrics, you can use construction paper or pages from magazines.
- You might solicit the help of a parent volunteer or two on days 2 and 3 to help with cutting and distributing material.
- You might make copies of student designs so that they can reference them when reassembling their pieces of material.
- You could offer a reward to the students who waste the least amount of material.

Name: _____ Date: _____

Grade 6, Lab #3: Crazy Quilt

In this lab you will work with a partner to explore area as it relates to quilting. You will:

1. decompose figures into familiar geometric shapes.
2. calculate the area of geometric figures using common area formulas.
3. plot geometric figures on the coordinate plane.
4. design your own crazy quilt square.

Day 1: Cozying up to Quilting Patterns

Crazy quilting refers to a type of quilt in which a crazy patchwork of materials is used to make the quilt. In crazy quilting, many different types of fabrics are used and a variety of embellishments are often included. This style of quilt became popular in the 1800s and is considered to be a type of textile folk art.

Part 1: Breaking it Down

The image your teacher gave you is one square in a crazy quilt. Each of the numbered pieces is made with a different kind of material. Use the quilt square provided by your teacher to answer the following questions.

1. Identify what type of polygon each piece is. Be as specific as possible.

    1. _____ 2. _____ 3. _____

    4. _____ 5. _____ 6. _____

    7. _____ 8. _____ 9. _____

    10. _____

2. For which of the polygons included in the quilt block do you know the area formulas? List those formulas.

    _____

    _____

    _____

    _____

    _____

    _____

3. For which of the polygons in the quilt block do you NOT know the area formulas? Draw lines through those polygons to break them down into shapes for which you DO know the area formulas.

Name: _____ Date: _____

Grade 6, Lab #3: Crazy Quilt

Day 1: Cozying up to Quilting Patterns

Part 2: How Much Material Was Used?

1. Quilters plan their squares carefully, making sure they have enough material for each piece in their design. Calculate how much material is needed to make the Crazy Quilt Square, using the pattern provided. Assume the quilt square is 32 centimeters by 32 centimeters and that each gridline marks off 1 centimeter.

1.	2.	3.
4.	5.	6.
7.	8.	9.
10.		

2. How do your answers compare to those of your classmates?

_____

_____

_____

_____

Name: _____ Date: _____

Grade 6, Lab #3: Crazy Quilt

Days 2 and 3: A Crazy Design of Your Own

Part 1: Design Time

1. Work with your partner to design a crazy quilt square. Your design must meet the following criteria:

   a. The pieces must fit together as a square that is 18 centimeters by 18 centimeters.

   b. Your square must have at least six different pieces.

   c. Your square may have no more than 10 different pieces.

   d. Your square must have at least two triangles.

   e. Your square must have at least one piece that is composed of more than one polygon.

2. Once you and your partner have decided on a design, use a straightedge to draw your design on the grid paper provided by your teacher, and label each of the pieces.

3. Plan which colors or materials you would like to use for each piece of your quilt, and record that on your pattern.

4. Have your teacher verify that your design meets all of the criteria.

Name: _____ Date: _____

Grade 6, Lab #3: Crazy Quilt

Days 2 and 3, Part 2: Make Your Crazy Quilt

1. Determine how much of each type of material you will need.

1.	2.	3.
4.	5.	6.
7.	8.	9.
10.		

2. Take your list and your design to your teacher, and ask for the amount of each type of material you calculated in step 1.

3. Cut the pieces of your design apart.

4. Pin the pieces to the appropriate material and cut them out.

5. Glue your material pieces onto your cardstock square.

6. Work with your classmates to assemble your class crazy quilt on the wall.

Name: _____ Date: _____

Grade 6, Lab #3: Crazy Quilt

Crazy Quilt Square Pattern

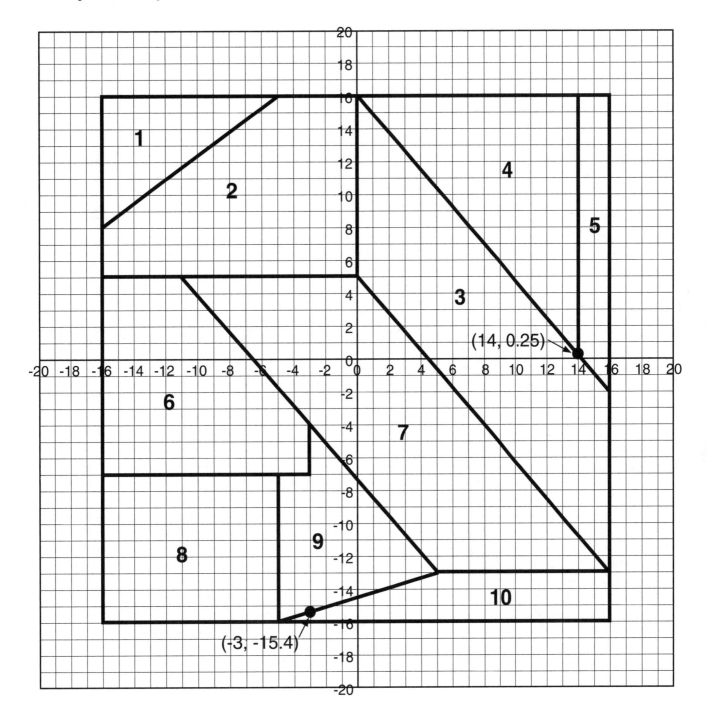

Name: _____   Date: _____

Grade 6, Lab #3: Crazy Quilt

Grid Paper for Crazy Quilt Design

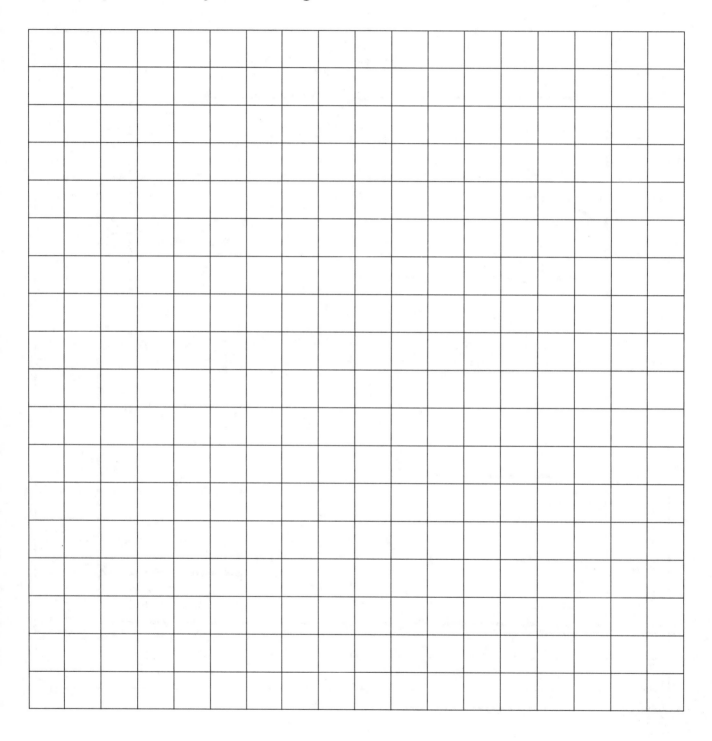

Grade 6, Lab #4: Road Trip

Teacher Information

Introduction

In this lab, students will learn more about ratios and proportional reasoning as they plan an imaginary road trip.

Duration of Lab

4 to 5 days

Common Core State Standards

- 6.RP.A.1
- 6.RP.A.2
- 6.RP.A.3
- 6.NS.B.2
- 6.NS.B.3

Prerequisite Skills

Before completing this activity, students need to be able to:
- add, subtract, multiply, and divide decimals.
- calculate ratios and unit rates.

Supplies

Each STUDENT will need...	Each GROUP will need...
<ul><li>Copy of the daily lab sheets</li><li>Copy of U.S. map</li><li>Two unlined index cards</li></ul>	<ul><li>Road atlas</li><li>Computer with Internet access</li></ul>

Teacher Tips

- This lab works best in pairs.
- Some students may not be aware of fun places to visit in the United States. You may wish to work as a class to create a list on the board before asking the students to plan their trip.
- If the students reside outside of the 48 contiguous states, have them select a "hometown" in the 48 contiguous states where they will begin and end the trip.
- You may wish to direct students to certain car rental and hotel websites.
- If your school computer security prevents students from accessing car rental and hotel websites, you will need to provide some of this information for the students.
- You may wish to divide Day 3 into three days with one dedicated to rental cars; one to accommodations, food, and entertainment; and one dedicated to revisions.
- All student answers for this activity will vary. If you would like to have some of the answers be the same, you can assign certain destinations and travel routes.

Name: _____ Date: _____

Grade 6, Lab #4: Road Trip

In this lab you will work with a partner to plan a
fictional road trip. You will:
1.   create an itinerary.
2.   calculate distances and time to travel.
3.   create a budget.

Day 1: Imagination Destination

  Taking a road trip across the United States can be a lot of fun! In this lab, we will pretend
that you and your partner are old enough to drive and travel without an adult. You are going to
work together to plan a two-week road trip.

Part 1: Do You Know Where You're Going?
1.   Locate your hometown on the U.S. map in the atlas.
2.   Spend some time looking at the U.S. map with your partner, and choose four places that
     you would like to visit on your road trip. The first place you choose needs to be in your
     home state. Keep in mind that a) this is a round trip…that is, you're going to come back to
     the place you started; b) you also need to remember that you will only be traveling for two
     weeks; c) you will travel at an average rate of 60 miles per hour and will travel no more
     than 7 hours in one day. List the four places you would like to visit.

_____

_____

3.   Using a pencil, mark your hometown and the four places you plan to visit on the blank U.S.
     map provided to you. Then sketch the general path you would like to travel.

Part 2: How Far Do We Have to Go?
1.   Turn to the page in your atlas that shows the map of your
     state. Locate your hometown on that map.
2.   Locate the scale of the map and then copy it onto the edge
     of an index card.
3.   Although roads are not straight, you can use your index
     card scale to determine the approximate distance between
     your hometown and your first destination.

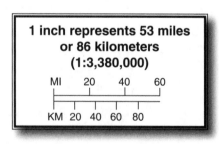

  Measure that distance and record it here: _____.

4.   If the atlas you are using has a mileage chart on the page, use it to refine your estimate

  and record it here: _____.

5.   Record your answer from #4 (or #3) on your U.S. map.
6.   Now, determine the rest of the distances between the places you plan to visit. Record
     these distances on your map.

Name: _____ Date: _____

Grade 6, Lab #4: Road Trip

Day 2: Over the River and Through the Woods

Part 1: Are We There Yet?

1. You have two weeks (14 days) for your road trip. Work with your partner to decide how long you would like to spend in each of your destination spots. Don't forget that you must leave some time to travel, so you may have to adjust this number later. Record that information in your destination itinerary table.

Destination Itinerary

	Destination	Number of Days	Number of Nights Spent in Hotel
1.			
2.			
3.			
4.			

2. Look at the distance between your hometown and your first destination. Assuming that you will be able to travel at an average rate of 60 miles per hour, calculate how long it will take you to reach your first destination. Record that information in your travel itinerary table.

3. Assuming that you do not want to drive more than seven hours per day, determine the number of days it will take you to travel to your first destination. Record that information in your travel itinerary table.

4. If it will take you more than one day to travel to your first destination, determine the number of nights you will need hotel accommodations. Record that information in your travel itinerary table.

5. Repeat steps 2 through 4 for your other destinations.

Travel Itinerary Table

	Destination	Number of Days	Travel Time to Destination (hours)	Number of Days Needed for Travel	Number of Travel Nights Spent in Hotel
1.					
2.					
3.					
4.					
5.	Home				

Name: _____ Date: _____

Grade 6, Lab #4: Road Trip

Days 3 and 4: Is This Going to Be Expensive?

You have a budget of $5,000 for your trip. This must cover all of the expenses for you and your partner for your entire trip. So choose your rental car, hotels, and activities wisely!

Part 1: Traveling by Car

1. Research the cost of rental cars online. You may choose to look directly at particular car rental companies or use a free online service that will search multiple rental companies at once to help you find a good deal. When selecting your car, be sure to select one from a rental company that gives you unlimited mileage and one that gets good gas mileage. Choose three cars you and your partner are interested in renting and record information about them in the table below.

Rental Company	Type of Car	Gas Mileage	Daily Rental Fee

2. Use the information in your table to decide what car to rent for your trip. Write a couple of sentences explaining which car you chose and why. _____

_____

_____

3. Assume an average cost of $3.91 per gallon of gas. Complete the "Rental Car" and "Gas" columns in your budget based on the car you chose.

Part 2: Room, Board, and Entertainment

1. Research the cost of hotels online. You may choose to look directly at particular hotel websites or use a free online service that will search multiple hotels at once to help you find a good deal. You will need to determine which hotels you will stay in along the way to and once you reach your destinations. Then fill that information into your budget.
2. Determine how much you are going to spend on food for you and your partner each day. You should budget at least $20 per person per day. Complete the "Food" column in your budget.
3. Research points of interest along your travel route and at your destinations. Determine which places you would like to visit and how much they will cost. Keep in mind that some places you visit will increase your food costs for that day. Complete the "Entertainment" column in your budget.

Part 3: Revise

1. Total your daily expenses in your budget. Then total your expenses for the entire trip.
2. If you are over budget, you will need to cut some of your expenses. Work with your partner to revise your budget. Remember that you must allot at least $20 per day per person for food and you cannot spend more than a total of $5,000. Submit your completed budget to your teacher with the map of your travel route.

Name: _____ Date: _____

Grade 6, Lab #4: Road Trip

Road Trip Budget

Day	Stopping Location	Rental Car	Gas	Food	Entertainment	Hotel	Daily Total
1							
2							
3							
4							
5							
6							
7							
8							
9							
10							
11							
12							
13							
14							
						TOTAL	

Name: _____ Date: _____

Grade 6, Lab #4: Road Trip

U.S. Map

Grade 7, Lab #1: Carnival Games

Teacher Information

Introduction

In this lab, students will explore probability concepts as they relate to games. Further, they will use their understanding of how probability is calculated and how areas of figures are determined to develop their own carnival game.

Duration of Lab

3 days

Common Core State Standards

- 7.G.B.4
- 7.G.B.6
- 7.NS.A.2
- 7.NS.A.3
- 7.SP.C.5
- 7.SP.C.6
- 7.SP.C.7
- 7.SP.C.8

Prerequisite Skills

Before completing this activity, students need to be able to:
- determine experimental probabilities.
- calculate the area of geometric figures.
- calculate percents.

Supplies

Each STUDENT will need...	Each GROUP will need...	The TEACHER will need...
• Copy of the daily lab sheets	• Wheel of Misfortune copied on card stock • Paper clip • Sharpened pencil • Ruler • 1 small paper plate	• Small paper plates • Yard stick • Masking tape • Marshmallows • Poster board • Markers

Teacher Tips

- This lab works best in pairs or small groups.
- The Wheel of Misfortune is divided into eighths and then four of those eighths are divided further—three of them into thirds and one into fourths.
- On Day 1, have two groups start with the Going Fishing! game.
- For one Going Fishing! game, place one paper plate in each tile on your floor in a 3-by-3 pattern as shown in the diagram. If you do not have a tiled floor, organize the plates in this same pattern in a 3-foot by 3-foot square. Use the masking tape to mark the starting line on the floor so that it is one yard away from the edge of the square. Students should stand behind this line to toss the marshmallows.

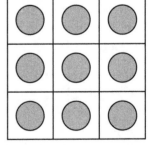

- You could also have students toss quarters, juggling balls, or some other object that doesn't bounce too much in the Going Fishing! game.

Name: _____ Date: _____

Grade 7, Lab #1: Carnival Games

In this lab you will work with a partner or in a small group to:
1. play carnival games and analyze the probability of winning.
2. create your own carnival games and analyze the probability of winning.

Day 1: Step Right Up!

From the crazy rides to the cotton candy to the games of chance, carnivals can be lots of fun! Have you ever wondered how carnival games are designed and why you win some more easily than others? You're going to analyze a couple of games in an attempt to answer this big question and more.

Part 1: Spin the Wheel of Misfortune

Carnival barkers often invite you to step right up and spin a wheel to win a prize. You and your partners are going to spin the Wheel of Misfortune and calculate the probability of being a winner.
1. Unfold one side of your paper clip so that it makes an "arrow" like the one shown in the picture.
2. Place the paper clip on the Wheel of Misfortune, and put the pencil through the paper clip so that the point is on the center of the wheel.
3. Flick the paper clip so that it spins around. Let each person in your group practice flicking the spinner. Then you can begin to collect some data.
4. Within your group, take turns spinning the paper clip a total of 50 times. When the paper clip stops moving, record whether it lands on a winning or losing spot. If the spinner lands on a line, spin again.

Spin	Win or Lose?	Spin	Win or Lose?	Spin	Win or Lose?	Spin	Win or Lose?	Spin	Win or Lose?
1		11		21		31		41	
2		12		22		32		42	
3		13		23		33		43	
4		14		24		34		44	
5		15		25		35		45	
6		16		26		36		46	
7		17		27		37		47	
8		18		28		38		48	
9		19		29		39		49	
10		20		30		40		50	

5. Based on your experiment, what is the probability of:
   a. winning this game? Write your answer as a fraction _____ and as a percent _____.
   b. losing this game? Write your answer as a fraction _____ and as a percent _____.

Name: _____ Date: _____

Grade 7, Lab #1: Carnival Games

Day 1: Step Right Up!
Part 2: Analyze the Wheel of Misfortune

1. Take a quick glance at the Wheel of Misfortune.

    a. Estimate the *percent* of the wheel that is shaded (lose). _____

    b. Estimate the *percent* of the wheel that is not shaded (win). _____

2. Now, look more closely at how the wheel is divided.

    a. What *fraction* of the wheel is shaded (lose)? _____

    b. What *fraction* of the wheel is not shaded (win)? _____

3. Calculate the area of the entire circle in square inches. Use 3.14 for π and round your answer to the nearest tenth of an inch. _____

4. Use your answers to questions 2 and 3 to calculate the area (to the nearest tenth of an inch) of the circle that is:

    a. shaded (lose). _____   b. not shaded (win). _____

5. To the nearest tenth, what percent of the Wheel of Misfortune:

    a. is shaded (lose)? _____   b. is not shaded (win)? _____

6. How do your answers to question 5 compare with your answers to question 5 in Part 1?

    _____

    _____

Part 3: Going Fishing!

Winning a goldfish at a carnival can be exciting! You and your partners are going to see how good you would be at a very similar game.

1. Stand behind the line and try to toss a marshmallow onto one of the plates. Let each person in your group practice tossing a marshmallow. Then you can begin to collect some data.

2. Within your group, take turns tossing a marshmallow a total of 50 times. Record whether it lands on a plate or on the floor. If it lands outside of the game area, toss it again.

Toss	Plate or Floor?	Toss	Plate or Floor?	Toss	Plate or Floor?	Toss	Plate or Floor?	Toss	Plate or Floor?
1		11		21		31		41	
2		12		22		32		42	
3		13		23		33		43	
4		14		24		34		44	
5		15		25		35		45	
6		16		26		36		46	
7		17		27		37		47	
8		18		28		38		48	
9		19		29		39		49	
10		20		30		40		50	

Name: _____   Date: _____

Grade 7, Lab #1: Carnival Games

Day 1, Part 3 (cont.)

3. Based on your experiment, what is the probability of :

    a. winning this game? Write your answer as a fraction _____ and as a percent _____.

    b. losing this game? Write your answer as a fraction _____ and as a percent _____.

Part 4: Analyze the Going Fishing! Game

1. Take a quick glance at the Going Fishing! Game.

    a. Estimate the *percent* of the game board that does not have a plate (lose). _____

    b. Estimate the *percent* of the game board that has a plate (win). _____

2. In square inches, calculate the *area* of:

    a. the entire game board. _____

    b. all of the plates. Use 3.14 for π and round to the nearest tenth. _____

3. Use your answers to question 2 to determine:

    a. the *fraction* of the game board that does not contain a plate (lose). _____

    b. the *fraction* of the game board that contains a plate (win). _____

4. To the nearest tenth, what percent of the game board:

    a. does not contain a plate (lose)? Round to the nearest tenth. _____

    b. contains a plate (win)? Round to the nearest tenth. _____

5. How do your answers to Question 4 compare with your answers to Question 3 in Part 3?

    _____

    _____

    _____

Part 5: What Determines a Good Carnival Game?

1. What similarities did you notice about these games? _____

    _____

2. What differences did you notice about these games? _____

    _____

3. There are a variety of factors involved in creating a carnival game. Brainstorm with your group and list at least four things you think are important when creating a carnival game.

    _____

    _____

    _____

Name: _____ Date: _____

Grade 7, Lab #1: Carnival Games

Day 2: 7th Grade Carnival

Part 1: Design Your Own Game
1. You and your partners need to design a carnival game that meets the following criteria:
    a. It must be an "area" game. That is, players must land or hit winning spaces with another object.
    b. The chance of losing must be slightly greater than the chance of winning.
    c. It should be fun!
2. Make a sketch of your game and use it to:

    a. calculate the area in which one can win. _____

    b. calculate the area in which one can lose. _____

    c. calculate the total area. _____

3. Use your answers above to predict the probability of winning and losing your game.

    _____

    _____

Part 2: Create Your Game
1. Use the materials in your classroom to create your game.

Day 3: Time to Play

Part 1: Let's Go to the Carnvial!
1. Set up your game.
2. Choose someone from your group to run the game and someone else to record the results.
3. Allow your classmates to play your game and record the number of wins and losses.
4. Use the results to calculate the experimental probabilities of winning and losing your

    game. _____

5. How do your answers to exercise 4 compare with your answers to exercise 3 from Day 2?

    _____

    _____

Name: _____ Date: _____

Grade 7, Lab #1: Carnival Games

Wheel of Misfortune

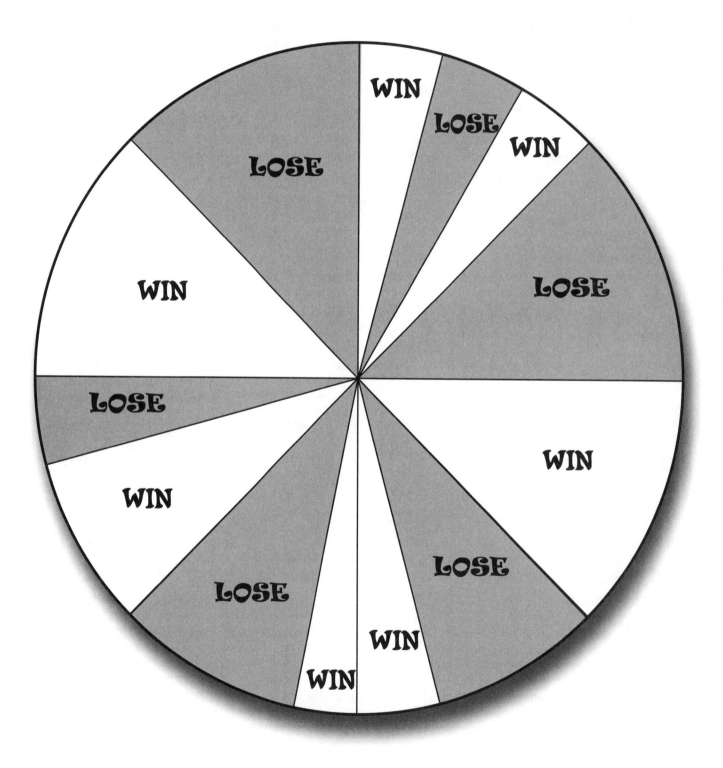

Grade 7, Lab #2: The Stroop Effect

Teacher Information

Introduction
In this lab, students will explore proportional relationships and statistics through data collected by administering a Stroop test.

Duration of Lab
2 days

Common Core State Standards
- 7.RP.A.2
- 7.NS.A.1
- 7.NS.A.2
- 7.NS.A.3
- 7.SP.A.1
- 7.SP.A.2
- 7.SP.B.3
- 7.SP.B.4

Prerequisite Skills
Before completing this activity, students need to be able to:
- write ratios to express the relationship between two quantities.
- use variables to represent variable quantities.
- plot points on a coordinate grid.
- calculate means.

Supplies

Each STUDENT will need...	Each GROUP will need...
• Copy of the daily lab sheets • Copy of Experiment Results table • Calculator • Graph paper • Straightedge	• Copy of Stroop lists • Tape • Scissors • Stop watch

Teacher Tips
- This lab works best in groups of four.
- Before students begin collecting data, assign each student a number from 1 to 30 to correspond to the numbers in the Experiment Results table.
- Although it may not be exact, there should be a relatively proportional relationship between incongruences and mean time.
- As students analyze the results of their experiment and discover that there is a proportional relationship between incongruences and mean time, help them make the connection between rates of change and the constant of proportionality.

Grade 7, Lab #2: The Stroop Effect

In this lab you will work with a partner or in a small group to:
1.   collect and analyze data related to the Stroop Effect.
2.   create graphs.
3.   determine whether or not two quantities have a proportional relationship.
4.   calculate the constant of proportionality.
5.   use data to draw inferences about a population.

Day 1: Name that Shape!

The Stroop Effect is named after psychologist John Ridley Stroop and is a demonstration of reaction time and how that changes when there is interference. Originally the Stroop Effect was demonstrated using color words. If a color word is printed in the same color the word names (that is, the word red is printed in red ink), it is easy to name the actual color in which the word is printed. If a color word is printed in a different color than the word names (that is, the word red is printed in blue ink), it is more difficult to name the actual color in which the word is printed.

In this lab, you will test your classmates' reaction times using shapes and shape words to demonstrate the Stroop Effect.

Part 1: Trial Preparation
1.   Cut apart the vertical lists of shapes with shape words provided by your teacher. Do NOT cut individual shapes apart.
2.   Tape both pieces of List 1 together so that you have one continuous List 1 and so that the header "List 1" is only visible at the top of the list.
3.   Repeat Step 2 with the other lists.
4.   Assign each person in your group one of the following roles: timer, recorder, subject, and scientist. The job description for each of these roles is given below:
     a.   Timer – Use the stopwatch to record each trial.
     b.   Recorder – Record the times for each subject.
     c.   Subject – Name each shape in the list as directed by the scientist.
     d.   Scientist – Direct the experiment, including telling the subject and the timer when to start and stop.

Part 2: Time to Experiment
1.   Complete one trial with each of the lists for a single subject. You are only recording the total time it takes the subject to name each shape in the list. You do *not* need to record the number of mistakes.
2.   Record the results of each trial in the Experiment Results table. Be sure to record the data in the correct row and remember to record the student's gender. Then each student in the group should add the results to their copy of the Experiment Results table.
3.   Assign new roles to each student in your group as listed in Part 1 Step 4.
4.   Repeat Steps 1 through 3 until every member of the group has had an opportunity to be the subject. With each new subject, be sure to begin with a different list to minimize the effect of practice. (Be sure to record the data in the column under the correct list.)

Name: _____ Date: _____

Grade 7, Lab #2: The Stroop Effect

Day 2: Analyze This!
Part 1: Get Up Close and Personal
1. Look at the class data in the Experiment Results table. List at least three observations you can make about the data.

   _____

   _____

2. Look at the lists of shapes and shape words and count the number of "incongruences" in each list. That is, count the number of times the shape word is *not* the same as the actual shape. Write that number at the top of the appropriate column in your Experiment Results table.

3. Do you think there is a relationship between the number of incongruences and the time needed to complete each list? Explain your reasoning.

   _____

   _____

Part 2: Time to Graph and Calculate
1. Calculate the mean time for each list for females, males, and all of the subjects. Record your answers in the Experiment Results table.
2. What do you notice about the mean times and the number of incongruences?

   _____

   _____

3. Graph the number of incongruences and mean times on a coordinate grid. Use different shapes to represent the female, male, and all subject data.
4. Based on your graph, do you think the incongruences and mean times have a proportional relationship? Explain your reasoning.

   _____

   _____

5. Calculate the rate of change between each pair of data points for the female data. Do these rates indicate a proportional relationship? Explain your reasoning.

   _____

   _____

6. Calculate the rate of change between each pair of data points for the male data. Do these rates indicate a proportional relationship? Explain your reasoning.

   _____

   _____

7. Calculate the rate of change between each pair of data points for the all subjects data. Do these rates indicate a proportional relationship? Explain your reasoning.

   _____

   _____

Name: _____ Date: _____

Grade 7, Lab #2: The Stroop Effect

Experiment Results

		Trial 1 (List 1)	Trial 2 (List 2)	Trial 3 (List 3)	Trial 4 (List 4)	Trial 5 (List 5)
Incongruences per List:						
Student	**Gender**	\multicolumn{5}{c}{Time for Each Trial}				
1						
2						
3						
4						
5						
6						
7						
8						
9						
10						
11						
12						
13						
14						
15						
16						
17						
18						
19						
20						
21						
22						
23						
24						
25						
26						
27						
28						
29						
30						
Mean time for females:						
Mean time for males:						
Mean time for all subjects:						

Grade 7, Lab #2: The Stroop Effect

List 1	List 2	List 3	List 4	List 5
square	triangle	square	square	pentagon
circle	rectangle	pentagon	triangle	square
square	circle	rectangle	rectangle	pentagon
circle	rectangle	pentagon	square	triangle
triangle	rectangle	triangle	rectangle	square
circle	circle	triangle	rectangle	circle
triangle	rectangle	pentagon	square	rectangle
rectangle	square	square	rectangle	triangle
pentagon	pentagon	rectangle	pentagon	circle

Grade 7, Lab #2: The Stroop Effect

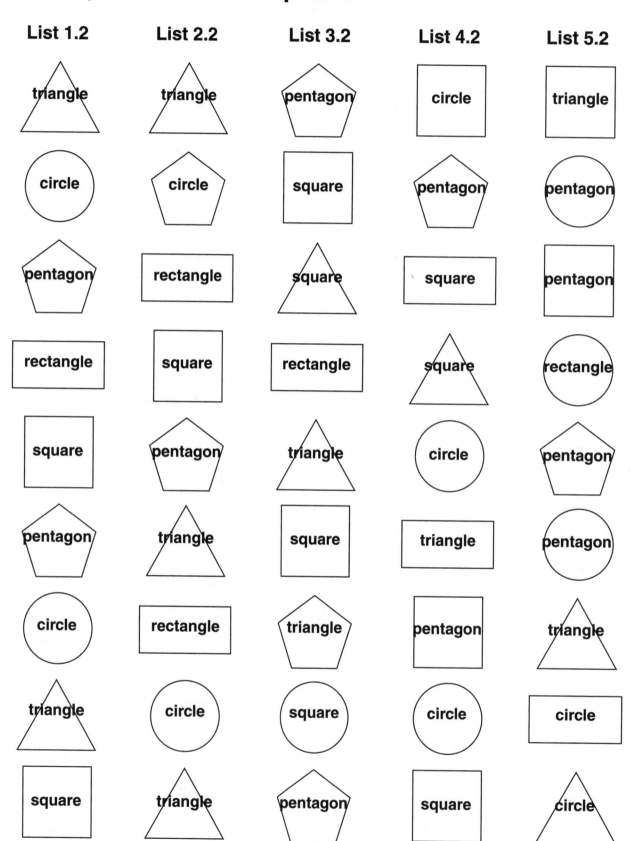

List 1.2	List 2.2	List 3.2	List 4.2	List 5.2
triangle	triangle	pentagon	circle	triangle
circle	circle	square	pentagon	pentagon
pentagon	rectangle	square	square	pentagon
rectangle	square	rectangle	square	rectangle
square	pentagon	triangle	circle	pentagon
pentagon	triangle	square	triangle	pentagon
circle	rectangle	triangle	pentagon	triangle
triangle	circle	square	circle	circle
square	triangle	pentagon	square	circle

Grade 7, Lab #3: Show Me the Money!

Teacher Information

Introduction
In this lab, students will explore proportional relationships and use formulas in Excel to analyze the growth of investments.

Duration of Lab
3 days

Common Core State Standards
- 7.RP.A.2
- 7.RP.A.3
- 7.EE.A.2
- 7.EE.B.3
- 7.EE.B.4

Prerequisite Skills
Before completing this activity, students need to be able to:
- define variables for variable quantities.
- express the relationships between quantities using an equation.
- write percents as decimals.

Supplies

Each STUDENT will need...	Each GROUP will need...
• Copy of the daily lab sheets	• 2 six-sided number cubes • A computer with Excel and Internet access • A calculator

Teacher Tips
- This lab works best in pairs.
- It will be helpful to your students if you have an understanding of simple interest and how it works.
- The Excel keystrokes provided for the activities on Days 2 and 3 are for Excel 2007. Prior to using this lab, work through the activities for Days 2 and 3 using the version of Excel that the students will use to make sure the keystrokes are the same.

Name: _____ Date: _____

Grade 7, Lab #3: Show Me the Money!

In this lab you will work with a partner to:
1. invest pretend money.
2. write and use formulas in Excel to help you make decisions about your investments.
3. write and use formulas in Excel to monitor your investments.

Day 1: It's Time to Invest!

Part 1: Calculating Simple Interest

   People can invest their money in a variety of ways. When money is invested, the hope is that interest will be made on that money. The simplest kind of investment yields simple interest. This is calculated based on the amount invested, the time for which it is invested, and the interest rate.

1. Let's say that you invest $200 in an account that has a 5% interest rate. How much do you think you would make in interest? Explain your reasoning.

   _____

   _____

   _____

2. Write a numerical expression to show how you can calculate 5% interest on $200.

   _____

3. The amount you invest is known as the principal. Let's say you are trying to decide how much principal to invest in an account that has an interest rate of 5%. Write an expression you could use to calculate the interest you would make. Use *P* to represent the principal.

   _____

4. Simple interest is always calculated on the principal amount. However, the time of the investment may vary and will affect the amount of interest earned. The table shows how much interest would be earned on $200 invested in an account that has an interest rate of 5% for two years, three years, four years, and five years. Record the amount you calculated in Step 1. Then explain the relationship between time in years and the amount of interest earned.

Time (years)	Interest
1	
2	$20
3	$30
4	$40
5	$50

   _____

   _____

   _____

   _____

   _____

Name: _____   Date: _____

Grade 7, Lab #3: Show Me the Money!

Day 1, Part 1 (cont.)

5. Write a numerical expression to show how you can calculate 5% interest on $200 over three years.

    _____

6. The amount of time you allow the principal to collect interest can vary. Revise the expression you wrote in Step 5 to include time. Use *t* to represent time.

    _____

7. So far, we have examined an account that returns 5% simple interest. However, interest rates vary from account to account. Revise the expression you wrote in Step 6 to replace the 5% interest rate with a variable. Use *r* to represent the interest rate.

    _____

8. Use your expression to calculate the interest earned on $200 invested in an account with a 5% interest rate for five years. Do your results match those in the table in Step 4? If not, revise your expression.

    _____

    _____

9. Research simple interest online. Write at least three sentences explaining what simple interest is and how it is calculated.

    _____

    _____

    _____

    _____

10. Based on your own work and your research, write an equation that can be used to calculate the simple interest for any principal amount, invested at any interest rate, for any amount of time.

    _____

Name: _____   Date: _____

Grade 7, Lab #3: Show Me the Money!

Day 2: Excelling at Investing

Part 1: Money to Invest

You are going to invest pretend money in a savings account that pays simple interest. You will use six-sided number cubes to determine how much you will invest and the amount of interest you will earn.

1. Roll both six-sided number cubes. Multiply the total value shown on the number cubes by 100 to determine how much you will invest. Write that amount here:

   _____

2. Roll both six-sided number cubes again to determine the interest rate for your investment. Let the smaller number represent the ones place and the larger number represent the tenths place. For example, if you roll a 3 and 5, then your interest rate will be 3.5%. Write that amount here:

   _____

3. Write an expression that can be used to determine how much interest you will earn given your principal (Step 1) and your interest rate (Step 2) over time (t). Write your expression in

   a. non-simplified form: _____

   b. simplified form: _____

Part 2: Using Excel to Monitor Investments

The amount of interest you earn depends on how long you leave your investment in the account. We can use Excel to help us monitor our investments so that we will know the best time to withdraw them and collect the interest.

1. Open a new Excel workbook. Type "Principal" in cell A1 and "Interest rate" in cell A2. Then, enter your principal from Part 1 Step 1 in cell B1 and your interest rate from Part 1 Step 2 as a decimal in cell B2.

2. In order to preserve your principal and interest rate to be used in later expressions, you must name them using these steps.

   a. Click on the principal amount that you entered in cell B1.

   b. Click on the formulas tab.

   c. Click on "Define name."

   d. Click on "Define name...", verify that the name says "Principal" and that the cell reference is correct. Then, click OK.

	A	B	C
1	Principal		
2	Interest Rate		
3			
4	Time (years)	Interest	Total
5	1		
6	2		
7	3		
8	4		

Name: _____ Date: _____

Grade 7, Lab #3: Show Me the Money!

Day 2, Part 2 (cont.)

> e. Repeat steps **a** through **d** for the interest rate, except use the value in B2 and name the value
> "Interest_rate."

3. Type "Time (years)" in cell A4, "Interest" in cell B4, and "Total" in cell C4. Then, save your file.

4. Enter 1, 2, 3, and so on for the number of years in column A until you get to 10.

5. You are going to enter your expression from Part 1 Step 3 into Excel. Do you think it is better to use the non-simplified or simplified expression? Explain your reasoning.

_____

_____

_____

_____

6. Regardless of your answer to Step 5, you are going to enter the non-simplified expression from Part 1 Step 3a. To do this, go to cell B5 and follow the steps below to enter your expression for calculating interest.
   a. Press the equal sign.
   b. Click on the "Formulas" tab.
   c. Click on "Use in formula" and then click "Principal."
   d. Type an asterisk to represent multiplication.
   e. Click on "Use in formula" and then click "Interest_rate."
   f. Type another asterisk to represent multiplication.
   g. Click on cell A5 for time. Your expression should look like this:
      =Principal*Interest_rate*A5.
   h. Press Enter. Cell B2 should now display the amount of interest earned on your principal in one year.

7. Verify that you entered your expression correctly by using the expression you wrote in Part 1 Step 3a. If you do not get the same answer, try entering your expression into cell B5 again.

8. Copy your expression to cells B6 through B14.

9. Verify that you have copied the expression correctly by using the expression you wrote in Part 1 Step 3a for year 5.

Name: _____ Date: _____

Grade 7, Lab #3: Show Me the Money!

Day 2, Part 2 (cont.)

10. How can you calculate the total amount of money you would have if you withdrew your investment after one year? Explain how you determined your answer.

    _____

    _____

    _____

    _____

11. Write an expression that you could use to determine the total amount of money you would have at the end of *t* years. Write your expression in

    a. non-simplified form: _____

    b. simplified form: _____

12. Enter either the non-simplified or simplified form of the expression in cell C5 in your Excel worksheet. Remember to begin your expression with an equal sign and to use the defined names for the principal. Verify that you have entered the expression correctly by calculating the total by hand.

13. Copy the expression into cells C6 through C14 to determine the total amount you could withdraw at the end of years 2 through 10.

14. Save your Excel worksheet.

Day 3: Do I Have Enough?

Part 1: Time and Money

When money earns simple interest, you have to be patient with your investments.

1. Open the Excel worksheet that you were working in yesterday.

2. Based on the data shown in your worksheet, how many years do you think you would need to leave your money in your account before your total was double the amount of your initial principal?

    _____

3. Extend your table to determine how many years it would take to double your initial principal

    and record the number of years here: _____

4. How close was your estimate in Step 2 to the actual value you calculated in Step 3?

    _____

    _____

Name: _____ Date: _____

Grade 7, Lab #3: Show Me the Money!

Day 3, Part 2: Using Excel to Monitor Investments

1. Let's say that you have decided to start saving for college and would like to save at least $5,000 in the next five years. Use your Excel worksheet to determine how much money (principal) you would have to put in a savings account that has an interest rate of 3.25% so that you would have a total of at least $5,000 in five years. How much would you have to invest? Explain how you determined your answer.

   _____

   _____

   _____

   _____

2. Let's say that you have decided to start saving for a cruise and would like to save at least $6,200. Use your Excel worksheet to determine how many years it would take you to earn at least $6,200 if you invest $2,500 in an account that has an interest rate of 4.3%. How many years would it take? Explain how you determined your answer.

   _____

   _____

   _____

   _____

3. Let's say that you have $482 and would like to invest it for seven years so that you can buy a car. You would like to have at least $2,000 for the car. Use your Excel worksheet to determine what interest rate your account would need to pay in order to have at least $2,000 in seven years.

   a. What interest would your account need to pay? Explain how you determined your answer.

   _____

   _____

   _____

   _____

   _____

   b. Do you think it is likely that you would be able to find an account that pays the interest rate needed to save this amount in seven years? Explain your reasoning.

   _____

   _____

   _____

Grade 7, Lab #4: The Great Pyramid of Spaghetti

Teacher Information

Introduction
In this lab, students will explore scale drawings, geometric shapes, and surface area.

Duration of Lab
4 days

Common Core State Standards
- 7.G.A.1
- 7.G.A.2
- 7.NS.A.2
- 7.NS.A.3

Prerequisite Skills
Before completing this activity, students need to be able to:
- calculate the area of triangles, rhombuses, and squares.
- recall the properties of equilateral triangles, rhombuses, and squares.
- plot points on a coordinate grid.
- add, subtract, multiply, and divide rational numbers.

Supplies

Each STUDENT will need...	Each GROUP will need...
• Copy of the daily lab sheets • Centimeter ruler • Protractor • Centimeter grid paper	• 30 pieces of dry spaghetti • 50 mini-marshmallows • Fine-tip permanent marker

Teacher Tips
- This lab works best in pairs or small groups.
- Because marshmallow work can be messy, you may wish to cover the students' work surfaces with plastic tablecloths or paper.
- Students will probably want to snack as they build. You may wish to provide extra marshmallows for them to snack on so that they do not eat the materials they need to build.

Name: _____ Date: _____

Grade 7, Lab #4: The Great Pyramid of Spaghetti

In this lab you will work with a partner or in a small group to:
1. make a scale drawing.
2. make a scale model.

Day 1: Pyramid Schema

From the Great Pyramids in Egypt to the Louvre Pyramid in Paris, there are many amazing pyramids in our world. In this lab, you and your classmates will add to the list of amazing pyramids by constructing your own.

The Louvre Pyramid in Paris is constructed of metal and glass and is located in the courtyard of the Louvre Museum in Paris. Three smaller pyramids of similar design surround it. The picture shows part of the large pyramid and one of the smaller pyramids.

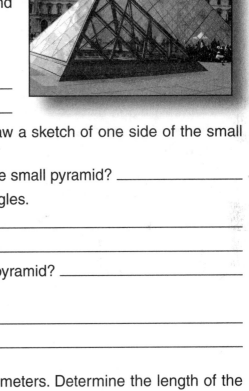

1. What shapes do you notice in the small pyramid?

    _____

    _____

2. On the back of this paper or on your own paper, draw a sketch of one side of the small pyramid.

3. How many equilateral triangles are on one side of the small pyramid? _____

4. List the side and angle properties of equilateral triangles.

    _____

    _____

5. How many rhombuses are on one side of the small pyramid? _____

6. List the side and angle properties of rhombuses.

    _____

    _____

7. The base of the small pyramid is approximately $\frac{70}{9}$ meters. Determine the length of the base of each of the equilateral triangles along the base of one side. Explain how you determined your answer.

    _____

    _____

    _____

8. What is the length of each side of the rhombuses and of the equilateral triangles on one side of the small pyramid? Explain how you determined your answer.

    _____

    _____

    _____

Grade 7, Lab #4: The Great Pyramid of Spaghetti

Day 2: Scaling it Down

1. You are going to make a scale drawing of one side of the small pyramid on the centimeter grid paper provided. Use a scale of $\frac{35}{72}$ meter to 1 centimeter and your answer to Day 1 Step 8 to determine the length of each side of each equilateral triangle and rhombus in your scale drawing. _____

2. Draw x- and y-axes on your grid paper so that they run along the base and left side of the paper as shown in the figure at the right.

3. Plot the point (0, 0). This will be the bottom left vertex of your scale drawing.

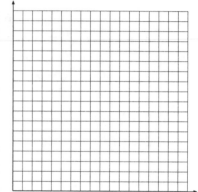

4. Use a centimeter ruler and a protractor to make a scale drawing of one side of the small pyramid on the grid. Mark and label each vertex. It is okay to estimate the coordinates when necessary.

Days 3 and 4: Can You Build It?

Part 1: Determining Your Scale

1. You and your partner are going to create a scale model of one of the small pyramids at the Louvre. In order to do this, you will need to determine the scale of your model. Measure one piece of spaghetti in centimeters.

    Write your answer here: _____

2. The piece of spaghetti will be the base of each side of your pyramid. Using the length of your piece of spaghetti and the actual length of the base of a small Louvre pyramid (Day 1 Step 7), determine your scale.

    _____

3. Determine the length of each side of each equilateral triangle and each rhombus in your scale model.

    _____

Name: _____ Date: _____

Grade 7, Lab #4: The Great Pyramid of Spaghetti

Days 3 and 4: Can You Build It?

Part 2: Preparing Your Materials

The construction of the pyramid works best when you use as many long pieces of dry spaghetti as possible. The spaghetti is relatively strong, but can also be broken easily. Use your centimeter ruler and permanent marker to mark off the following pieces.

1. 8 full-length pieces of spaghetti
2. 12 half-length pieces of spaghetti
3. 24 quarter-length pieces of spaghetti

Then, carefully break the pieces where they have been marked. The closer you hold the spaghetti to the markings when breaking it, the more precise your pieces will be.

Part 3: Constructing Your Pyramid

1. Take each of the full-length pieces of spaghetti and string three mini-marshmallows on them so that the pieces are divided into fourths.

2. Take each of the half-length pieces of spaghetti and string one mini-marshmallow on them so that the pieces are divided in half.

3. Lay four of the full-length pieces in a rectangle on the table and use mini-marshmallows to connect them at the corners.

4. Take the remaining four full-length pieces and push them into each of the corners and then connect them with one mini-marshmallow at the top.

5. Take one quarter-length piece and use it to connect the second marshmallow from the left on the base of one side to the second marshmallow from the bottom on the edge of the side to its left.

6. Take one quarter-length piece and use it to connect the second marshmallow from the right on the base of one side to the second marshmallow from the bottom on the edge of the side to its right.

7. Repeat Steps 5 and 6 on each of the other sides. See Figure 1.

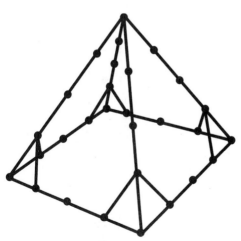

Figure 1

Name: _____ Date: _____

Grade 7, Lab #4: The Great Pyramid of Spaghetti

Days 3 and 4, Part 3 (cont.)

8. Take one half-length piece and use it to connect the middle marshmallow on the base of one side to the middle marshmallow on the edge to its right.

9. Take one half-length piece and use it to connect the middle marshmallow on the base of one side to the middle marshmallow on the edge to its left.

10. Repeat Steps 8 and 9 on each of the other sides. See Figure 2.

11. Use the remaining quarter-length and half-length pieces to complete each side. Refer to your scale drawing for the placement of equilateral triangles and rhombuses.

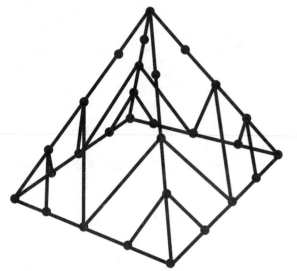

Figure 2

Part 4: Reflecting on Your Pyramid

Write a short paragraph (at least three sentences) describing how scale drawings and models might be helpful when working on a large-scale construction project.

_____

_____

_____

_____

_____

_____

_____

_____

Name: _____ Date: _____

Grade 7, Lab #4: The Great Pyramid of Spaghetti

Grid Paper for Pyramid Scale Drawing

Grade 8, Lab #1: Crime Scene Investigation

Teacher Information

Introduction
In this lab, students will collect data, create and analyze scatter plots, model data with lines, and use linear models to make predictions.

Duration of Lab
3 days

Common Core State Standards
- 8.EE.B.5
- 8.EE.C.8
- 8.F.A.1
- 8.F.A.2
- 8.F.A.3
- 8.F.B.4
- 8.SP.A.1
- 8.SP.A.2
- 8.SP.A.3

Prerequisite Skills
Before completing this activity, students need to be able to:
- measure lengths to the nearest tenth of a centimeter using a ruler or tape measure.
- graph points on a coordinate grid.
- evaluate an equation for a given value of one of the variables.
- write an equation using point-slope formula.

Supplies

Each STUDENT will need...	Each GROUP will need...
• A copy of the student pages • A copy of class data collection sheet • A copy of crime scene ulna sketch • A straightedge	• A centimeter tape measure • A yard stick

Teacher Tips
- This lab works best in pairs or small groups.
- Explain that standard measurements (such as inches) are only used in the United States and that the rest of the world uses metric measurements. That is why crime scene investigators would use centimeters for their measurements.
- You may help students decide on an appropriate and consistent scale to use when creating scatter plots.
- You may wish to review guidelines for drawing a best-fit line.

Name: _____ Date: _____

Grade 8, Lab #1: Crime Scene Investigation

Forensic scientists examine evidence found at crime scenes in an effort to make connections that might help solve the crime. In this lab, you will pretend you are a forensic scientist who found a human ulna at a crime scene. You have three main goals:

1. Determine the relationship between the length of the ulna and a person's height.
2. Use a mathematical relationship between the length of the ulna and height to predict the height of a person.
3. Identify the person to whom the ulna belonged given physical characteristics of that person.

Day 1: How Do You Measure Up?

Part 1: Measuring Body Parts

For this part of the lab, you will work with a partner.

1. Use the yard stick to measure your partner's height. Then have your partner measure your height.

    My height in inches: _____

2. Convert your heights to centimeters. Round your answers to the nearest tenth. (HINT: 1 inch ≈ 2.54 centimeters)

    My height in centimeters: _____

3. Record your heights in centimeters on the class data collection sheet.

4. Use the tape measure to measure your partner's ulna (from the elbow to the wrist) to the nearest tenth of a centimeter. Then have your partner measure your ulna.

    Length of my ulna in centimeters: _____

5. Record your results on the class data collection sheet.

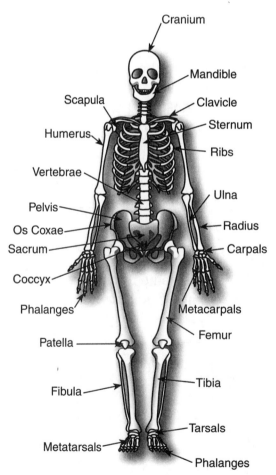

Cranium

Mandible

Scapula

Clavicle

Sternum

Humerus

Ribs

Vertebrae

Ulna

Pelvis

Radius

Os Coxae

Carpals

Sacrum

Coccyx

Metacarpals

Phalanges

Femur

Patella

Tibia

Fibula

Tarsals

Metatarsals

Phalanges

Name: _____ Date: _____

Grade 8, Lab #1: Crime Scene Investigation

Day 1: How Do You Measure Up?

Part 2: Informal Analysis of the Data

Review the data on the class data collection sheet. Then answer the following questions

1. How do the female and male ulna measurements compare to each other?

    _____

    _____

2. How do the female and male heights compare to each other?

    _____

    _____

3. Suppose the human ulna that was found at the crime scene was 27.5 centimeters long.

    a. Would you expect that it had belonged to a male or female? Explain how you deter-
        mined your answer.

        _____

        _____

        _____

    b. Estimate the height of the person to whom the ulna belonged. Explain how you deter-
        mined your answer.

        _____

        _____

        _____

Name: _____ Date: _____

Grade 8, Lab #1: Crime Scene Investigation

Day 2: Feeling Scattered?

Part 1: Create a Scatter Plot of the Data

Use the information from the class data collection sheet to complete the following activity.

1. Create a scatter plot that compares the length of the ulnas with the heights of the females in your class.

2. Describe the relationship between the ulna lengths and heights of the females.

   _____

   _____

   _____

   _____

   _____

   _____

   _____

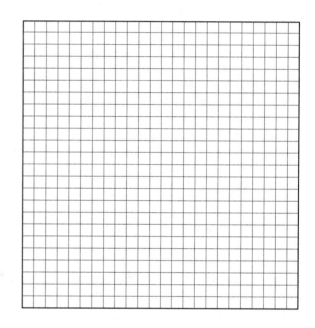

3. Create a scatter plot that compares the length of the ulnas with the heights of the males in your class.

4. Describe the relationship between the ulna lengths and heights of the males.

   _____

   _____

   _____

   _____

   _____

   _____

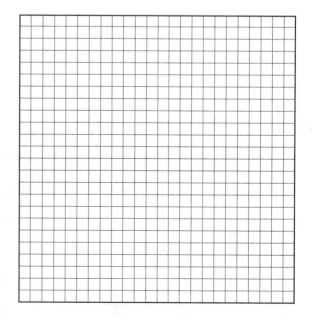

Name: _____ Date: _____

Grade 8, Lab #1: Crime Scene Investigation

Day 2: Feeling Scattered?

Part 2: Draw a Best-Fit Line

1. Use a straightedge to draw a best-fit line on each of your graphs.

2. Explain how you determined where to place your line on each graph.

   _____

   _____

Part 3: Formally Analyze the Data

1. Suppose the human ulna that was found at the crime scene was 27.5 centimeters long.
   a. Using your best-fit lines, would you expect that it had belonged to a male or female? Explain how you determined your answer.

   _____

   _____

   b. Using your best-fit lines, estimate the height of the person to whom the ulna belonged. Explain how you determined your answer.

   _____

   _____

2. The equation $y = 5.66x + 32.16$ can be used to predict the height of a female in centimeters (y) for a given ulna length (x). Suppose the human ulna that was found at the crime scene was 27.5 centimeters long and that you know that it belonged to a female. Use the equation to approximate her height. Show your work.

   _____

3. If you predicted that the ulna belonged to a female in Question 1a, then how does your answer to Question 2 compare to your answer to Question 1b?

   _____

   _____

4. The equation $y = 17.45x - 254.03$ can be used to predict the height of a male in centimeters (y) for a given ulna length (x). Suppose the human ulna that was found at the crime scene was 27.5 centimeters long and that you know that it belonged to a male. Use the equation to determine how tall he must have been. Show your work.

   _____

5. If you predicted the ulna belonged to a male in Question 1a, then how does your answer to Question 4 compare to your answer to Question 1b?

   _____

   _____

Name: _____ Date: _____

Grade 8, Lab #1: Crime Scene Investigation

Day 2, Part 3 (cont.)

6. You can write an equation for each of your best-fit lines using the point-slope formula.

    a. Write an equation for the line representing the female data. How does your equation compare to the equation given in Question 2?

    _____

    _____

    b. Write an equation for the line representing the male data. How does your equation compare to the equation given in Question 4?

    _____

    _____

    c. Do your lines intersect? If so, what does that point represent in the context of this problem? If not, what does that mean in the context of this problem?

    _____

    _____

    _____

    _____

Day 3: Solve the Crime

1. Measure the sketch on page 65 that shows the ulna found at the crime scene. Record the measurement to the nearest tenth of a centimeter.

    _____

2. There are four missing persons in your case load. They are described below. Using your answer to Question 1 and the data you collected, determine which missing person you think this ulna belonged to. Show your calculations and include an explanation to support your reasoning.
    *   Jane Doe – Female who was 157.5 centimeters tall.
    *   Janet Doe – Female who was 165.1 centimeters tall.
    *   John Doe – Male who was 177.8 centimeters tall.
    *   Jim Doe – Male who was 172.7 centimeters tall

    _____

    _____

    _____

    _____

Name: _____ Date: _____

Grade 8, Lab #1: Crime Scene Investigation

Class Data Collection Sheet

Females

Name	Ulna (cm)	Height (cm)

Males

Name	Ulna (cm)	Height (cm)

Grade 8, Lab #1: Crime Scene Investigation

Sketch of Ulna Found at Crime Scene

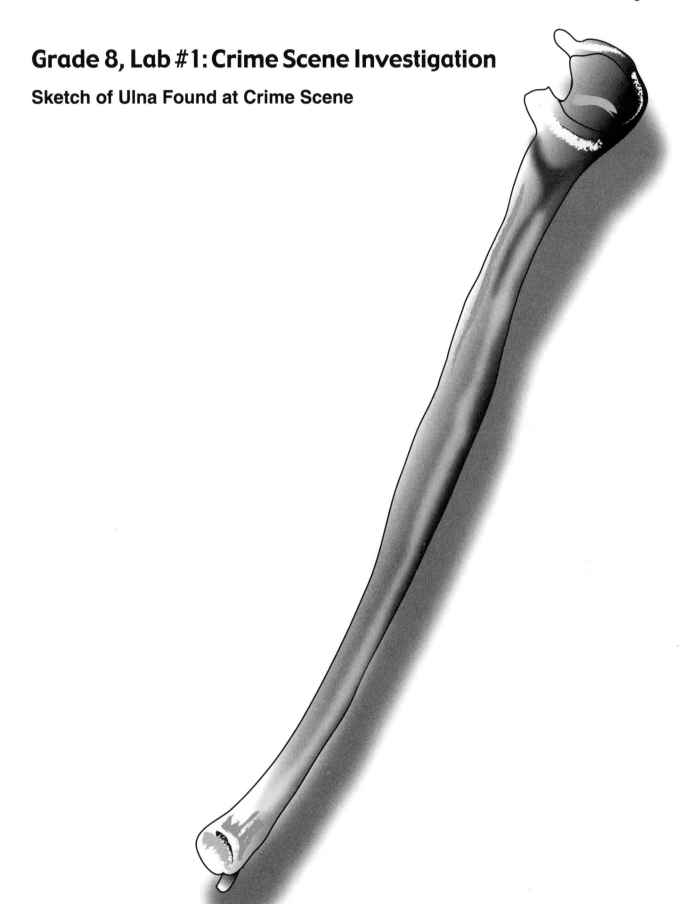

Grade 8, Lab #2: Pennsylvania Dutch Hex Signs

Teacher Information

Introduction
In this lab, students will explore geometric shapes, transformations, and the Pythagorean Theorem through Pennsylvania Dutch hex signs.

Duration of Lab
2 days

Common Core State Standards
- 8.G.A.2
- 8.G.A.3
- 8.G.A.4
- 8.G.B.7
- 8.G.B.8
- 8.NS.A.1
- 8.NS.A.2
- 8.EE.A.2

Prerequisite Skills
Before completing this activity, students need to be able to:
- identify different types of triangles.
- use the Pythagorean Theorem.
- understand translations, rotations, reflections, and dilations.
- determine whether figures are similar or congruent.

Supplies

Each STUDENT will need...	Each GROUP will need...
• Copy of the daily lab sheets • Tracing paper • Straightedge	• Colored pencils, markers, or crayons • A computer with Internet access to look up hex sign designs (optional)

Teacher Tips
- This lab works best in pairs.
- "Patty paper" (the paper used in between hamburger patties) makes excellent tracing paper and can be found at restaurant supply stores.
- You may wish to print pictures of hex designs or check out books from the library so that students have examples to look at when creating their own designs.
- The hex sign that is provided is rather complicated. Encourage students to make simple hex designs that meet the criteria outlined.

Name: _____ Date: _____

Grade 8, Lab #2: Pennsylvania Dutch Hex Signs

In this lab you will work with a partner to:
1. analyze a Pennsylvania Dutch hex sign.
2. create your own hex sign.

Day 1: Understanding Folk Art

If you drive through the countryside in eastern Pennsylvania, you will notice beautiful works of art in the form of Pennsylvania Dutch hex signs painted on the sides of barns. Some say that these designs are painted on the barns to protect the animals or to bring good luck. Whatever the reason for painting them, they have become valuable pieces of folk art, and many of them use geometric figures.

Part 1: Casual Analysis of a Hex Sign

1. Look at the given hex sign and list all of the different shapes you see. Be as specific as possible.

    _____

    _____

2. Look at triangles *ABC* and *ADC*. Do you think they are congruent? Explain your reasoning.

    _____

    _____

3. Use a piece of tracing paper and a straightedge to trace triangle *ABC*. Use the traced image to determine all of the triangles that are congruent to triangle *ABC* and list them below.

    _____

    _____

4. Use a piece of tracing paper and a straightedge to determine any other congruent triangles in the hex sign. Then list them below.

    _____

    _____

    _____

5. There are three circles in the hex sign. Are they similar? Explain your reasoning.

    _____

    _____

    _____

Name: _____ Date: _____

Grade 8, Lab #2: Pennsylvania Dutch Hex Signs

Day 1, Part 2: Mathematical Analysis of a Hex Sign

1. Look at triangle *ABC*. What kind of triangle do you think it is? Explain your reasoning.

    _____

    _____

2. Use mathematics to determine whether or not your answer in Step 1 is true. Show your work and explain how you determined your answer. (NOTE: The coordinates of the points pro-vided have been rounded. So, you should round your results to the nearest whole number.)

    _____

    _____

    _____

    _____

3. How could triangle *ABC* be transformed to create triangle *ADC*? Explain your reasoning.

    _____

    _____

4. How could triangle *ABC* be transformed to create triangle *FDE*? Explain your reasoning.

    _____

    _____

5. Look at triangle *CUV*. What kind of triangle do you think it is? Explain your reasoning.

    _____

    _____

6. Given the length of $\overline{CU}$ is about 3.72 units, use mathematics to determine whether or not your answer in Step 5 is true. Show your work and explain how you determined your an-swer. (NOTE: The coordinates of the points provided have been rounded. So, you should round your results to the nearest whole number.)

    _____

    _____

    _____

    _____

7. How could triangle *CUV* be transformed to create triangle *LγΒ*? Explain your reasoning.

    _____

    _____

Name: _____ Date: _____

Grade 8, Lab #2: Pennsylvania Dutch Hex Signs

Day 1, Part 2 (cont.)

8. How could triangle *CUV* be transformed to create triangle *KαZ*? Explain your reasoning.

    _____

    _____

9. Look at the circle with diameter $\overline{SK}$. How can it be transformed to create the circle with diameter $\overline{RH}$? Explain how you determined your answer.

    _____

    _____

    _____

Day 2: Creating Folk Art

Part 1: Design Your Hex Sign

   You and your partner will work together to design your own hex sign. Your design must meet the following criteria.

   a. It must have at least four congruent triangles.
   b. It must have one figure that is a dilation of another figure.
   c. It must have one figure that is formed by rotating another figure.
   d. It must have one figure that is a reflection of another figure.
   e. It must fit within a circle with an eight-centimeter radius.

1. Make a sketch of your design.

Part 2: Analyze and Create Your Hex Sign
   1. Verify that your design follows the criteria outlined above.
   2. Create your design on the provided coordinate grid.
   3. Color your design.
   4. On your own paper, explain which part of your design meets each part of the criteria, and provide mathematical support for your explanations.

Name: _____ Date: _____

Grade 8, Lab #2: Pennsylvania Dutch Hex Signs

Hex Sign Pattern

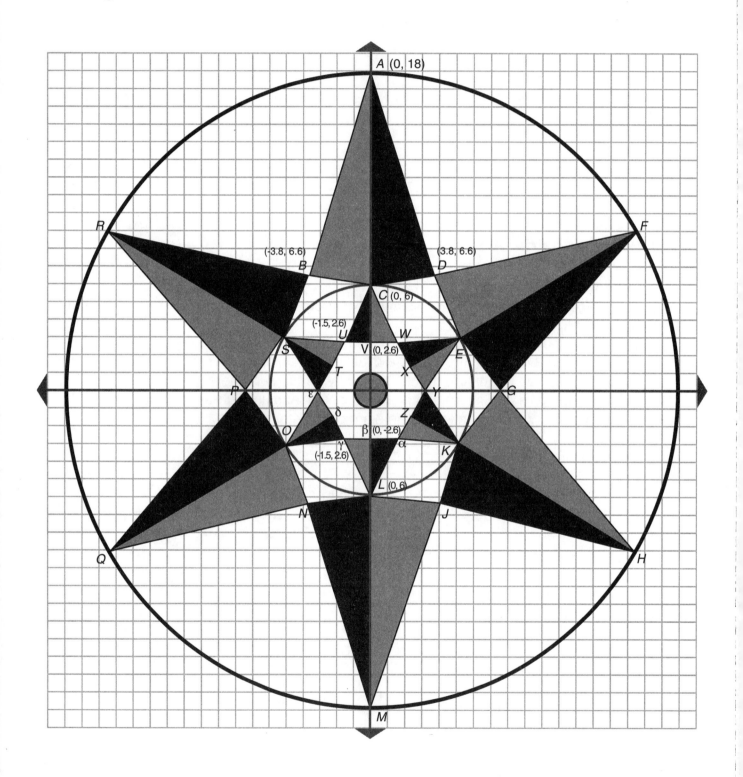

Name: _____ Date: _____

Grade 8, Lab #2: Pennsylvania Dutch Hex Signs

Grid Paper for Hex Sign

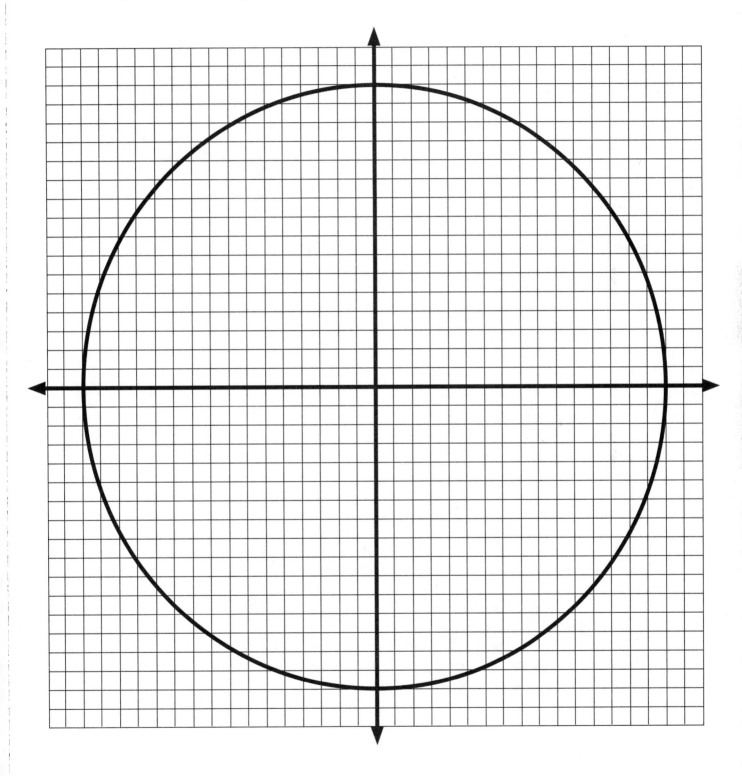

Grade 8, Lab #3: Strange, But True

Teacher Information

Introduction
In this lab, students will explore scientific notation.

Duration of Lab
2 days

Common Core State Standards
- 8.EE.A.1
- 8.EE.A.3
- 8.EE.A.4

Prerequisite Skills
Before completing this activity, students need to be able to:
- write large numbers using scientific notation.
- add, subtract, multiply, and divide numbers written in scientific notation.
- apply the properties of integer exponents.

Supplies

Each STUDENT will need...	Each GROUP will need...
• Copy of the daily lab sheets	• Computer with Internet access • Calculator

Teacher Tips
- This lab works best in pairs.
- Day 1 should be completed without the use of a calculator.
- After students complete Day 1 Part 1, you may wish to engage the class in a discussion about their answers to Step 6.
- Students should be able to find the answers to Day 2 Part 1 easily online. However, prior to giving this lab to your students, you may wish to make sure your school's security measures do not prevent students from finding the answers they need.
- Some of the data students collect will be given as a range. You may wish to have students record both numbers or choose just one.

Name: _____   Date: _____

Grade 8, Lab #3: Strange, But True

In this lab you will work with a partner to:
1. explore numbers written in scientific notation.
2. operate on numbers written in scientific notation.
3. research and analyze strange, but true numbers.

8.5 X 10⁹

Day 1: Something Smells

Part 1: Scientific Notation

Scientific notation can be very helpful when dealing with exceptionally large numbers or exceptionally small numbers. We can use it to make computation easier and save time.

1. Without using a calculator, determine the product of

    a. 900, 40, and 2,000. _____

    b. 9, 4, and 2. Then multiply the result by 1,000,000. _____

    _____

2. How does your answer in Step 1a compare with your answer in Step 1b?

    _____

    _____

3. Was it easier to calculate the product in Step 1a or Step 1b? Explain your reasoning.

    _____

    _____

4. Without using a calculator, determine the sum of

    a. 900, 40, and 2,000. _____

    b. 9, 4, and 2. _____

5. How does your answer in Step 4a compare with your answer in Step 4b?

    _____

    _____

6. Discuss with your partner the difference between the calculations in Steps 1 and 4.

    _____

    _____

Name: _____ Date: _____

Grade 8, Lab #3: Strange, But True

Day 1, Part 2: Scientific Notation and Stinky Feet

1. Have you ever wondered why feet stink? Research online to learn why feet stink. Write a short paragraph explaining why feet stink.

   _____

   _____

   _____

   _____

   Then write the number of sweat glands a person has in their feet here: _____

2. Calculate the number of feet sweat glands in your classroom. Explain how you determined your answer.

   _____

   _____

3. Would scientific notation be helpful in calculating the number of feet sweat glands in your classroom? Explain your reasoning.

   _____

   _____

4. There are approximately 3.18×10^8 people in the United States. Write this number in standard form.

   _____

5. How would you determine the total number of feet sweat glands on people in the United States?

   _____

   _____

6. Use your answer to Step 1 to write the number of sweat glands on the feet of one person in scientific notation.

   _____

7. Solve the two multiplication problems written below.

   a. 318,000,000 x 250,000 _____

   b. $(3.18 \times 2.5) \times (10^8 \times 10^5)$ _____

8. How does your answer in Step 7a compare with your answer in Step 7b?

   _____

   _____

Name: _____ Date: _____

Grade 8, Lab #3: Strange, But True

Day 1, Part 2 (cont.)

9. Was it easier to calculate the answer for Step 7a or the answer for Step 7b? Explain your reasoning.

    _____

    _____

10. There are approximately 1.75×10^{15} feet sweat glands in the world. Write this number in standard form.

    _____

11. How could you use the information about the number of feet sweat glands in the world to determine the total number of people in the world?

    _____

    _____

12. Solve the two division problems written below.

    a. $\dfrac{1{,}750{,}000{,}000{,}000{,}000}{250{,}000}$ _____

    b. $\dfrac{1.75}{2.5} \times \dfrac{10^{15}}{10^{5}}$ _____

13. How does your answer in Step 12a compare with your answer in Step 12b?

    _____

    _____

14. Was it easier to calculate the answer for Step 12a or the answer for Step 12b? Explain your reasoning.

    _____

    _____

    _____

Name: _____ Date: _____

Grade 8, Lab #3: Strange, But True

Day 2: Those Are Some Crazy Numbers!

Part 1: Researching the Facts

Use the Internet to find the answers to the following questions. In all cases, write your answers in both standard form and in scientific notation.

Question	Standard Form	Scientific Notation
1. How much does the Earth weigh in tons?		
2. How much does the United States spend on clothes each day?		
3. How many words are there in the English language?		
4. How many miles does the Earth travel in five trips around the sun?		
5. How many cats are in the United States?		
6. How many dogs are in the United States?		
7. How many gallons of saliva does a person produce in their lifetime?		

Name: _____ Date: _____

Grade 8, Lab #3: Strange, But True

Day 2, Part 2: Applying the Research

Use the data you collected in Day 2, Part 1 to answer the following questions. Compute each answer using numbers in scientific notation and show your work.

1. There are 2.0×10^3 pounds in 1 ton. How many pounds does the Earth weigh?

   _____

2. How much does the United States spend on clothes in one year?

   _____

3. If there are 3.18×10^8 people in the United States, how much does the average person spend on clothes each year?

   _____

4. If the average United States high school graduate's vocabulary is 6×10^4 words, how many times more words are there in the English language than in the graduate's vocabulary?

   _____

5. How many miles have you traveled around the sun?

   _____

6. How many cats and dogs are there in the United States?

   _____

7. How many more cats than dogs are there in the United States?

   _____

8. How many gallons of saliva will the people in the United States produce over their lifetimes?

   _____

Grade 8, Lab #4: Round and Round We Go!

Teacher Information

Introduction

In this lab, students will explore the relationship between the circumference and diameter of a circle and the relationship between the volumes of a cylinder, cone, and sphere with the same radius and height.

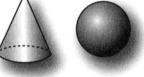

Duration of Lab

3 days

Common Core State Standards

- 8.G.C.9
- 8.NS.A.1
- 8.NS.A.2

Prerequisite Skills

Before completing this activity, students need to be able to:
- use the formula for calculating the circumference of a circle.
- use the formula for calculating the volume of a rectangular prism.
- solve an equation for a variable in terms of another variable.

Supplies

Each STUDENT will need...	Each GROUP will need...	The TEACHER will need...
• Copy of the daily lab sheets	• String • Centimeter and inch ruler • Copy of the cone net • Piece of 8.5 x 11-in. paper • Scissors • Tape • One round balloon • Newspaper • Paper maché glue • Mini-marshmallows or popped popcorn • Calculator	• A number of relatively flat circular objects of various sizes • Straight pins

Teacher Tips

- This lab works best in pairs.
- You may wish to cover student workspace on Day 2 when students create the paper maché sphere.
- Mix two parts white glue to one part water to make paper maché glue for Day 2.
- Deriving the formula for the volume of a sphere may be difficult for students. You may need to ask facilitating questions to help them through this valuable learning process.

Name: _____ Date: _____

Grade 8, Lab #4: Round and Round We Go!

In this lab you will work with a partner to:
1. discover where pi comes from.
2. discover the relationship between the volumes of a cylinder, cone, and sphere with the same radius and height.

Day 1: Who Needs Pi?

Part 1: Circular Measurements

Pi, or π, is a number that shows up in many mathematical formulas and equations that are related to circle figures or objects. Did you ever wonder why?

1. Get two circular objects from your teacher.
2. Use your piece of string to measure the circumference of a circular object. Lay the string along your ruler to determine the approximate circumference of the object in centimeters and record it in the table.
3. Use your piece of string to measure the diameter of the same object. Lay the string along your ruler to determine the approximate diameter of the object in centimeters and record it in the table.
4. Repeat Steps 2 and 3 for the other circular object.
5. Trade your objects with the objects of a group sitting near you. Then repeat Steps 2 and 3 with your new objects.
6. Repeat Step 5 three more times so that you have measured the circumference and diameter of a total of 10 objects.

Object	Circumference (cm)	Diameter (cm)	$\frac{C}{d}$

Part 2: What Does π Really Represent?
1. Use your calculator to determine the decimal approximation for the ratio of the circumference to diameter of each of your objects and record it in the table above. Round your answers to the nearest hundredth.
2. What do you notice about the ratio of circumference to diameter for your objects?

_____

3. Write the formula for calculating the circumference of a circle given the diameter. Then rewrite the formula by dividing both sides by d and simplifying.

_____

4. Write one sentence explaining what π represents.

_____

Name: _____ Date: _____

Grade 8, Lab #4: Round and Round We Go!

Day 2: Circular Construction

Part 1: Making a Cylinder
1. Fold the 8.5 x 11-in. piece of paper in half (hamburger fold) and then tear along the fold.
2. Roll one-half of the piece of paper to form a cylinder so that the two shorter sides are touching.
3. Tape the shorter sides together. (Do not overlap the edges.)
4. Place the cylinder on the other half-piece of paper and trace the base.
5. Cut the base out and tape it to the bottom of the cylinder.
6. How tall is the cylinder in inches? Write your answer here: _____
7. What is the circumference of the base of the cylinder in inches? Write your answer here: _____
8. Use the circumference formula to approximate the diameter of the base of the cylinder in inches. Use 3.14 for π, and round your answer to the nearest hundredth. Write your answer here: _____

Part 2: Making a Cone
1. Cut out the net of the lateral sides of the cone.
2. Roll the net into a cone and tape the edges together. (Do not overlap the edges.)
3. How tall is the cone in inches? Write your answer here: _____
4. Use a piece of string and an inch ruler to determine the circumference of the base of the cone. Write your answer here: _____
5. Use the circumference formula to approximate the diameter in inches of the base of the cone. Use 3.14 for π, and round your answer to the nearest hundredth. Write your answer here: _____

Part 3: Making a Sphere
The great circle of a sphere is the circle formed when a plane intersects the sphere such that it passes through the center of the sphere. The equator is the great circle of the Earth. You are going to make a sphere whose great circle has a measure of about 8.5 inches.
1. Blow up the balloon and pinch the opening closed. Have your partner use the string to measure the great circle. If it is 8.5 inches, tie off your balloon. If it is not 8.5 inches, add or let out air until you have a great circle of 8.5 inches.
2. Use the circumference formula to approximate the diameter of the great circle of the sphere. Use 3.14 for π, and round your answer to the nearest hundredth. Write your answer here: _____
3. Tear the newspaper into strips.
4. Dip the strips into the glue one at a time and lay them on the balloon.
5. Repeat Step 4 until the balloon is covered, except for an area of about two inches in diameter near where the balloon was tied off.
6. Set your paper maché balloon in a safe place to dry.

Name: _____ Date: _____

Grade 8, Lab #4: Round and Round We Go!

Day 3: Round and Round We Go!

Part 1: Estimating the Volume of a Cone

1. Collect your cylinder and cone.
2. You are going to use your cone to fill the cylinder with mini-marshmallows. How many times do you think you need to fill the cone with mini-marshmallows in order to have enough to fill the cylinder?

   _____

3. Test your hypothesis from Step 3.

   a. How many times did you have to fill the cone in order to fill the cylinder?

   _____

   b. Were you surprised by the results?

   _____

4. What is the relationship between the heights of the cylinder and the cone?

   _____

5. What is the relationship between the bases of the cylinder and the cone?

   _____

6. Based on your answers to Steps 3–5, write a statement describing the relationship between the volumes of a cylinder and a cone.

   _____

   _____

7. What is the general formula for the volume of a cylinder?

   _____

8. How should you modify the general formula for a cylinder to have a formula for the volume of a cone? Explain how you determined your answer.

   _____

   _____

Part 2: Estimating the Volume of a Sphere

1. Collect your sphere.
2. Use a straight pin to pop the balloon in your sphere, and then remove the deflated balloon.
3. You are going to use your cone to fill the sphere with mini-marshmallows. How many times do you think you need to fill the cone with mini-marshmallows in order to have enough to fill the sphere?

   _____

Name: _____ Date: _____

Grade 8, Lab #4: Round and Round We Go!

Day 3, Part 2 (cont.)

4. Test your hypothesis from Step 3.

    a. How many times did you have to fill the cone
        in order to fill the sphere?

        _____

    b. Were you surprised by the results?

        _____

5. What is the relationship between the diameters of the cone and the sphere?

    _____

6. What is the radius of the base of the cone and the sphere?

    _____

7. Divide the height of the cone by the radius and round your answer to the nearest whole
    number. What do you notice?

    _____

    _____

8. Use your answer to Step 7 to rewrite the general formula for the volume of a cone in terms
    of the radius.

    _____

    _____

    _____

9. Based on your answers to Steps 5–8, write a statement describing the relationship
    between the volume of a cone and a sphere.

    _____

    _____

    _____

10. How should you modify the general formula from Step 8 to have a general formula for the
    volume of a sphere? Explain how you determined your answer.

    _____

    _____

    _____

Grade 8, Lab #4: Round and Round We Go!

Net for Cone

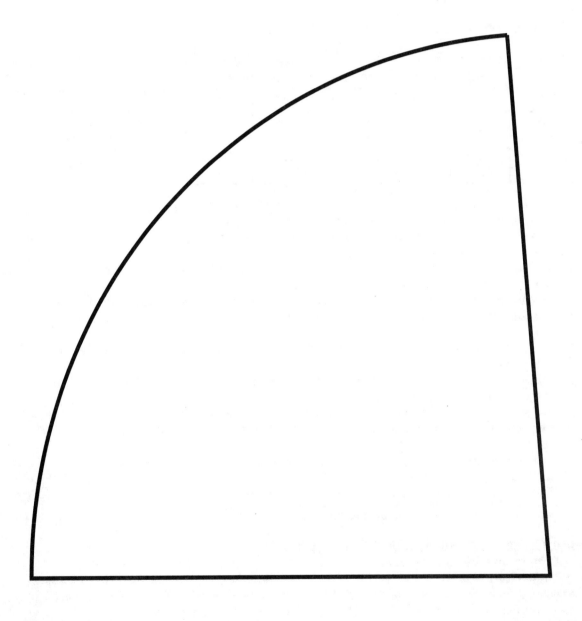

Answer Keys

Only definite answers are given. All other answers may vary.

Grade 6, Lab #1: Packing and Wrapping (pgs. 4–12)
Day 1: Fill it Up
Part 2: Determine Which Package Will Hold the Most
2. Answers will vary. However, students should indicate that they will need to determine the volume of the package.

Day 2: All Wrapped Up
Part 1: Determine Which Package Can Be Wrapped With the Least Amount of Paper
2. Answers will vary. However, students should indicate that they will need to determine the surface area of the packages.

Day 3: Short and Wide or Tall and Narrow?

1.

Package	Length (cm)	Width (cm)	Height (cm)	Surface Area (cm²)	Volume (cm³)
A	$3\frac{2}{5}$	$3\frac{1}{10}$	$2\frac{1}{2}$	$53\frac{29}{50}$	$26\frac{7}{20}$
B	$6\frac{4}{5}$	$6\frac{1}{5}$	5	$214\frac{8}{25}$	$210\frac{4}{5}$
C	3	2	3	42	18
D	6	4	6	168	144

What students label as *length, width,* and *height* may vary. As long as the three correct measurements are used for each package, the answers are acceptable.

2. a.   The dimensions of Package B are double those of Package A.
   b.   The dimensions of Package D are double those of Package C.
3. a.   The surface area of Package B is four times the surface area of Package A. The ratio of the surface area of Package A to the surface area of Package B is 1:4.
   b.   The surface area of Package D is four times the surface area of Package C. The ratio of the surface area of Package C to the surface area of Package D is 1:4.
4. Sample answer: If one rectangular prism has dimensions that are twice that of another rectangular prism, then the surface area of the larger prism will be four times the surface area of the smaller prism.
5. $y = 4x$
6. a.   The volume of Package B is eight times the volume of Package A. The ratio of the volume of Package A to the volume of Package B is 1:8.
   b.   The volume of Package D is eight times the volume of Package C. The ratio of the volume of Package C to the volume of Package D is 1:8.
7. Sample answer: If one rectangular prism has dimensions that are twice that of another rectangular prism, then the volume of the larger prism will be eight times the volume of the smaller prism.
8. $b = 8a$
9. a.   The surface area of Package E is 4 x 64, or 256 square inches, because doubling the dimensions of the rectangular prism quadruples the surface area.
   b.   The volume of Package F is 576 ÷ 8, or 72 cubic inches, because halving the dimensions of a rectangular prism reduces the volume by a factor of $\frac{1}{8}$.

Grade 6, Lab #2: It's Freezing Around Here! (pgs. 14–19)
Day 1: Getting Up Close and Personal
Part 1: Grise Fiord, Nunavut, Canada
1. Answers will vary but may include: 1) All of the temperatures are less than 0 degrees Celsius.; 2) May 3rd was the coldest day of the month.; 3) May 26th was the warmest day of the month.
2. The highest temperature was −1.8 degrees Celsius. It is the highest temperature because it is closest to 0 degrees Celsius.
3. The lowest temperature was −16.2 degrees Celsius. It is the lowest temperature because it is farthest from 0 degrees Celsius.

4.

The range of temperatures is approximately 14 degrees. I determined this by rounding the lowest and highest temperatures to the nearest whole number. Then, I calculated the difference of the absolute values of those numbers.

Part 2: Toronto, Ontario, Canada
1. Answers will vary but may include: 1) All of the temperatures are greater than 0 degrees Celsius.; 2) May 17th was the coldest day of the month.; 3) May 27th was the warmest day of the month.
2. The highest temperature was 24.3 degrees Celsius. It is the highest temperature because it is farthest from 0 degrees Celsius.
3. The lowest temperature was 8.0 degrees Celsius. It is the lowest temperature because it is closest to 0 degrees Celsius.
4.

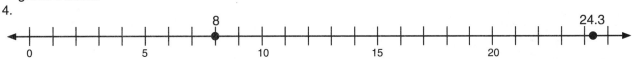

The range of temperatures is approximately 16 degrees. I determined this by rounding the highest temperature to the nearest whole number. Then, I calculated the difference of the highest and lowest temperatures.

Part 3: Comparing Temperatures
1. May 1st; The temperature in Grise Fiord was –10.8 degrees Celsius when it was 10.8 degrees Celsius in Toronto. Those numbers are the same distance from 0 on the number line but in opposite directions.
2. Yes, the two cities had opposite temperatures on May 1st/May 6th, May 4th/May 8th, May 8th/May 17th, and May 12th/May 7th and May 16th.
3. In May it was colder in Grise Fiord, where the temperatures were all below 0 degrees Celsius, than in Toronto, where the temperatures were all above 0 degrees Celsius.

Day 2: A Summary of the Cold Spots
Part 1: Dot Plots of the Cold Spots
3. Sample answer; These dot plots should be placed one above the other on the table, floor, or wall with the zeroes aligned.

Grise Fiord Temperatures for May 2014

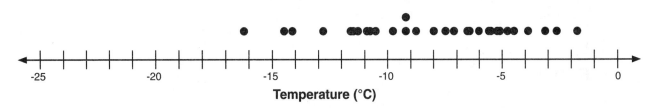

Toronto Temperatures for May 2014

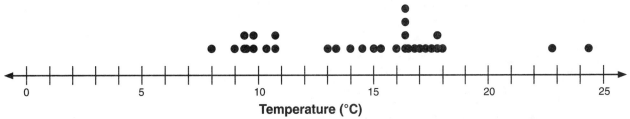

4. Answers will vary.
5. a. Sample answer: The temperatures are fairly evenly distributed and the distribution is relatively symmetric.
   b. Sample answer: The temperatures are skewed to the left with a couple of outliers in the twenties.
   c. Sample answer: The temperatures in Grise Fiord are much colder than those in Toronto.

Part 2: The Cold, Hard Facts
1. Because the distribution is relatively symmetric, I would choose mean and mean absolute deviation.
2. Mean: −8.1; Mean absolute deviation: 3.1
3. The average temperature for May 2014 in Grise Fiord is −8.1 degrees Celsius, and the daily temperatures varied from the mean an average of 3.1 degrees.
4. Because the distribution is skewed and has outliers, I would choose median and interquartile range.
5. Median: 15.3; Interquartile range: 6.7
6. The median temperature for May 2014 in Toronto is 15.3 degrees Celsius, and the middle 50 percent has a range of about 6.7 degrees.

Day 3: How's the Weather At Home?
Answers will vary.

Grade 6, Lab #3: Crazy Quilt (pgs. 21–26)
Day 1: Cozying up to Quilting Patterns
Part 1: Breaking it Down

1.
1. right triangle	2. pentagon	3. parallelogram
4. right triangle	5. trapezoid	6. pentagon
7. parallelogram	8. rectangle	9. pentagon
10. trapezoid		

2. Right triangle: $A = \frac{1}{2}bh$; rectangle: $A = bh$; parallelogram: $A = bh$; trapezoid: $A = \frac{1}{2}h(b_1 + b_2)$

3. Sample answer: Dashed lines are drawn to make shapes that have known area formulas.

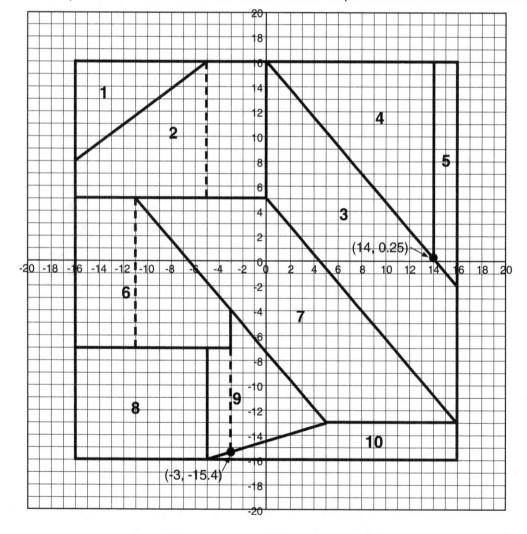

Part 2: How Much Material Was Used?
1.

1. $A = \frac{1}{2}bh$ $= \frac{1}{2}(11)(8)$ $= 44$ cm^2	2. $A = \frac{1}{2}h(b_1 + b_2) + bh$ $= \frac{1}{2}(11)(11 + 3) + (11)(5)$ $= 132$ cm^2	3. $A = bh$ $= (11)(16)$ $= 176$ cm^2
4. $A = \frac{1}{2}bh$ $= \frac{1}{2}(14)(15.75)$ $= 110.25$ cm^2	5. $A = \frac{1}{2}h(b_1 + b_2)$ $= \frac{1}{2}(2)(18 + 15.75)$ $= 33.75$ cm^2	6. $A = bh + \frac{1}{2}h(b_1 + b_2)$ $= (12)(5) + \frac{1}{2}(8)(12 + 3)$ $= 120$ cm^2
7. $A = bh$ $= (11)(18)$ $= 198$ cm^2	8. $A = bh$ $= (11)(9)$ $= 99$ cm^2	9. $A = \frac{1}{2}h(b_1 + b_2) + \frac{1}{2}bh$ $= \frac{1}{2}(2)(9 + 8.4) + \frac{1}{2}(11.4)(8)$ $= 63$ cm^2
10. $A = \frac{1}{2}h(b_1 + b_2)$ $= \frac{1}{2}(3)(21 + 11)$ $= 48$ cm^2		

2. The students' answers should all be the same. If they find differences, they should identify the errors made and correct them.

Grade 6, Lab #4: Road Trip (pgs. 28–32)
All answers for this activity will vary.

Grade 7, Lab #1: Carnival Games (pgs. 34–38)
Day 1: Step Right Up!
Part 1: Spin the Wheel of Misfortune
Answers will vary.
Part 2: Analyze the Wheel of Misfortune
1. a.   Answers may vary but should be close to 50%.
   b.   Answers may vary but should be close to 50%.

2 a. $\frac{1}{8} + \left(\frac{1}{3} \times \frac{1}{8}\right) + \frac{1}{8} + \left(\frac{2}{3} \times \frac{1}{8}\right) + \left(\frac{3}{4} \times \frac{1}{8}\right) + \left(\frac{1}{3} \times \frac{1}{8}\right) = \frac{49}{96}$

   b. $\frac{1}{8} + \left(\frac{1}{3} \times \frac{1}{8}\right) + \left(\frac{1}{3} \times \frac{1}{8}\right) + \frac{1}{8} + \left(\frac{1}{3} \times \frac{1}{8}\right) + \left(\frac{1}{4} \times \frac{1}{8}\right) + \left(\frac{2}{3} \times \frac{1}{8}\right) = \frac{47}{96}$

3. $A = \pi r^2$
    $= \pi(3.5)^2$
    ≈ 38.5 in.^{2}
4. a. $\frac{49}{96}(38.5) = 19.7$ in.^{2}
   b. $\frac{47}{96}(38.5) = 18.8$ in.^{2}
5. a. $\left(\frac{19.7}{38.5}\right) \times 100 = 51.2\%$
   b. $\left(\frac{18.8}{38.5}\right) \times 100 = 48.8\%$
6. Answers will vary.
Part 3: Going Fishing!
2. Answers will vary.
3. a.   Answers will vary, but if 8.5-inch diameter luncheon plates and 1-foot squares were used, it will be around $\frac{1}{3}$ or 33%.

   b.   Answers will vary, but if 8.5-inch diameter luncheon plates and 1-foot squares were used, it will be around
        $\frac{2}{3}$ or 67%.
Part 4: Analyze the Going Fishing! Game
   1. a.   Answers will vary, but if 8.5-inch diameter luncheon plates and 1-foot squares were used, it will be around
           33%.
      b.   Answers will vary, but if 8.5-inch diameter luncheon plates and 1-foot squares were used, it will be around
           67%.
   2. a.   36 x 36 = 1,296 in.^{2}
      b.   Sample answers for 2b., 3., and 4. using 8.5-inch diameter luncheon plates:
           $A = 9\pi r^2$
              $= 9(3.14)(4.25)^2$
              $= 510.4$ in.^{2}

   3. a. $\dfrac{1,296 - 510.4}{1,296} = \dfrac{785.6}{1,296}$

      b. $\dfrac{510.4}{1,296}$

   4. a. $\dfrac{785.6}{1,296}$ x 100 $\approx$ 60.6%

      b. $\dfrac{510.4}{1,296}$ x 10 $\approx$ 39.4%

   5. Answers will vary.
Part 5: What Determines a Good Carnival Game?
   1. Sample answer: Both of the games were "area" games, and I needed to be able to calculate the area of a
      circle. Both of the games required that something "land" on a winning space in order for me to win.
   2. Sample answer: I had a better chance of winning the Wheel of Misfortune than the Going Fishing! game. The
      Going Fishing! game required more skill than the Wheel of Misfortune, and the "losing" area was significantly
      greater than in the Wheel of Misfortune.
   3. Answers will vary.

Grade 7, Lab #2: The Stroop Effect (pgs. 40–42)
Day 1: Name that Shape!
The results of each trial of the experiment will vary by subject. However, the times should increase as students
complete each subsequent list (1 through 5).
Day 2: Analyze This!
Part 1: Get Up Close and Personal
   1. Answers will vary.
   2. List 1: 0 incongruences; List 2: 3 incongruences; List 3: 6 incongruences; List 4: 9 incongruences; List 5: 12
      incongruences
   3. Answers will vary. However, students should recognize that as the number of incongruences increases, so
      does the time.
Part 2: Time to Graph and Calculate
   2. Answers will vary. However, students should recognize that as the number of incongruences increases, so
      does the time.

Grade 7, Lab #3: Show Me the Money! (pgs. 46–51)
Day 1: It's Time to Invest!
Part 1: Calculating Simple Interest
   1. Sample answer: I would earn 200 x 0.05 or 10 dollars in interest, because 10 is 5% of 200.
   2. 200 x 0.05
   3. P x 0.05

4.

Time (years)	Interest
1	$10
2	$20
3	$30
4	$40
5	$50

Sample answer: Because you earn $10 each year, you can multiply the number of years by 10 to determine how much interest you earn.

5. Sample answer: 200 x 0.05 x 3 or 10 x 3
6. Sample answer: 200 x 0.05 x *t* or 10 x *t*
7. 200 x *r* x *t*
8. 200 x *r* x *t* = 200 x 0.05 x 5 = 50
9. Sample answer: Simple interest is interest that is paid on the principal amount invested. It is not paid on the interest that is accrued. The amount of interest is determined by the interest rate and the amount of time the principal is invested. It can be calculated by multiplying the principal by the interest rate in decimal form and then by time.
10. *I* = *Prt*

Day 3: Do I Have Enough?
Part 2: Using Excel to Monitor Investments
1. Sample answer: I would have to invest around $4,305 in order to have at least $5,000 in five years. I calculated this by entering 3.25% as my interest rate in the spreadsheet and then trying different amounts as the principal until I found an amount that would have a total of at least $5,000 in year 5.
2. Sample answer: It would take about 35 years for my investment to reach $6,200. I calculated this by entering $2,500 as my principal and 4.3% as my interest rate in the spreadsheet and then extending the years until I reached a total of $6,200.
3. a. Sample answer: I would have to invest in an account that had an interest rate of at least 45% to have at least $2,000 in seven years. I calculated this by entering $482 as my principal in the spreadsheet and then trying different interest rates until I found a rate that would make my total investment at least $2,000 in year 7.
   b. Sample answer: It is unlikely that I would find a savings account that would have an interest rate of 45%.

Grade 7, Lab #4: The Great Pyramid of Sphagetti! (pgs. 53–57)
Day 1: Pyramid Schema
1. Equilateral triangles and rhombuses
2.

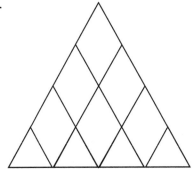

3. 4
4. All sides are congruent. All angles have a measure of 60°.
5. 6
6. All sides are congruent. The opposite angles are congruent and the adjacent angles are supplementary.
7. The length of the base of each triangle is $\frac{35}{18}$ meters. I divided the total length of the base, $\frac{70}{9}$ meters, by 4 because there are four equilateral triangles along the base.
8. The length of each side of the rhombuses and of the equilateral triangles on one side of the small pyramid is $\frac{35}{18}$ meters. Because the triangles are equilateral and the triangles and rhombuses share sides, I know that each side has a length of $\frac{35}{18}$ meters.

Day 2: Scaling it Down

1. Each side length is $\frac{35}{18} \div \frac{35}{72}$ or 4 centimeters.

4.

Days 3 and 4: Can You Build It?
Part 1: Determining Your Scale

1. Sample answer: 26 cm

2. The scale is 1 centimeter equals $\frac{35}{117}$ meter.

$$\frac{70}{9} \text{ m} = 26 \text{ cm} \longrightarrow \left(\frac{70}{9} \div 26\right) \text{ m} = 1 \text{ cm} \longrightarrow \frac{35}{117} \text{ m} = 1 \text{ cm}$$

3. Each side length is 6.5 centimeters.

$$\frac{35}{18} \text{ m} \div \frac{35}{117} \text{ m/cm} = \frac{35}{18}\text{m} \times \frac{117}{35} \text{ m/cm}$$
$$= 6.5 \text{ cm}$$

Grade 8, Lab #1: Crime Scene Investigation (pgs. 59–65)
Day 2: Feeling Scattered?
Part 1: Create a Scatter Plot of the Data

2. Answers may vary. However, the ulna length and height should have a positive linear association. That is, as the ulna length increases, so does the height.

4. Answers may vary. However, the ulna length and height should have a positive linear association. That is, as the ulna length increases, so does the height.

Part 2: Draw a Best-Fit Line

2. Answers will vary. However, students should indicate that they drew the line so that it followed the general pattern of the data with about the same number of data points above and below the line.

Part 3: Formally Analyze the Data

2. $y = 5.66x + 32.16$
   $= 5.66(27.5) + 32.16$
   $= 187.81$
   The female would have been about 187.8 cm tall.

4. $y = 17.45x - 254.03$
   $= 17.45(27.5) - 254.03$
   $= 225.845$
   The male would have been about 225.8 cm tall.

6. c. The lines may or may not intersect. If they do, the point of intersection represents the ulna length for which the height is the same for males and females. If they do *not* intersect, that means that there is no ulna length for which the height of males and females is the same.

Day 3: Solve the Crime

1. The length of the ulna is 24.8 cm.

2. Answers will vary. Students may choose to use their data and scatter plots, best-fit lines, or the equations provided to determine the person to whom the ulna belonged. The calculations for the equations are provided. These equations indicate that John Doe is the most likely candidate.

   Female:
   $y = 5.66x + 32.16$
   $= 5.66(24.8) + 32.16$
   $= 172.5$

   Male:
   $y = 17.45x - 254.03$
   $= 17.45(24.8) - 254.03$
   $= 178.7$

Grade 8, Lab #2: Pennsylvania Dutch Hex Signs (pgs. 67–71)
Day 1: Understanding Folk Art
Part 1: Casual Analysis of a Hex Sign

1. Circles, isosceles triangles, right triangles, kites, and a hexagon.

2. Answers will vary.

3. $\triangle ABC \cong \triangle ADC \cong \triangle FDE \cong \triangle FGE \cong \triangle HGK \cong \triangle HJK \cong \triangle MJL \cong \triangle MNL \cong \triangle QNO \cong \triangle QPO \cong \triangle RPS \cong \triangle RBS$

4. $\triangle CUV \cong \triangle CWV \cong \triangle EWX \cong \triangle EYX \cong \triangle KYZ \cong \triangle K\alpha Z \cong \triangle L\alpha\beta \cong \triangle L\gamma\beta \cong \triangle O\gamma\delta \cong \triangle O\varepsilon\delta \cong \triangle S\varepsilon T \cong \triangle SUT$
   $\triangle CUW \cong \triangle EWY \cong \triangle KY\alpha \cong \triangle L\alpha\gamma \cong \triangle O\gamma\varepsilon \cong \triangle S\varepsilon U$

5. Yes, because all circles are the same shape but can vary in size, all circles are similar.

Part 2: Mathematical Analysis of a Hex Sign

1. It is an isosceles triangle because two sides appear to be the same length.

2. I can determine the lengths of sides $\overline{AB}$ and $\overline{AC}$. Because points A and C have the same x-coordinates, I subtract their y-coordinates to determine the length of $\overline{AC}$. So, $AC = 18 - 6$ or 12. Then, I can use the Pythagorean Theorem to determine the length of $\overline{AB}$.
   $AB^2 = (3.8)^2 + (18 - 6.6)^2$
   $AB^2 = 14.44 + 129.96$
   $\sqrt{AB^2} = \sqrt{144.4}$
   $AB \approx 12$
   Because $\overline{AB}$ and $\overline{AC}$ are the same length, I know that $\triangle ABC$ is an isosceles triangle.

3. It could be reflected over the y-axis. The y-coordinates of points B and D are the same, but the x-coordinates are opposites.

4. It could be rotated about the origin because the triangles are the same size but have different orientations.

5. It is a right triangle because two sides appear to meet to form a right angle.

6. I can determine the lengths of $\overline{CV}$ and $\overline{UV}$. Because points C and V have the same x-coordinates, I subtract their y-coordinates to determine the length of $\overline{CV}$. So, $CV = 6 - 2.6$ or 3.4. Because points U and V have the same y-coordinates, I subtract their x-coordinates to determine the length of $\overline{UV}$. So, $UV = 0 - (-1.5)$ or 1.5.

Then, I can use the Pythagorean Theorem to determine whether or not $\triangle CUV$ is a right triangle.

$$CV^2 + UV^2 = CU^2$$
$$(3.4)^2 + (1.5)^2 \stackrel{?}{\approx} (3.72)^2$$
$$11.56 + 2.25 \stackrel{?}{\approx} 13.8$$
$$13.81 \approx 13.8$$

So, $\triangle CUV$ is a right triangle.

7. It could be reflected over the x-axis. The x-coordinates of the vertices in triangle CUV are the same as the x-coordinates in triangle $L\gamma\beta$, but the y-coordinates are opposites.

8. It could be reflected over the y-axis and then rotated about the origin because the triangles are the same size but have different orientations.

9. The circle with diameter $\overline{SK}$ has a diameter of 12 units, and the circle with diameter $\overline{RH}$ has a diameter of 36 units. Therefore, the circle with diameter $\overline{SK}$ could be dilated by a factor of three to form the circle with diameter $\overline{RH}$.

Day 2: Creating Folk Art
Part 2: Analyze and Create Your Hex Sign

4. Answers will vary. However, students should use ordered pairs and the Pythagorean Theorem to support their answers where appropriate.

Grade 8, Lab #3: Strange, But True (pgs. 73–78)
Day 1: Something Smells
Part 1: Scientific Notation

1. a. $900 \times 40 \times 2{,}000 = 72{,}000{,}000$
   b. $9 \times 4 \times 2 = 72$; $72 \times 1{,}000{,}000 = 72{,}000{,}000$
2. They are the same.
3. Answers will vary.
4. a. $900 + 40 + 2{,}000 = 2{,}940$
   b. $9 + 4 + 2 = 15$
5. They are very different.
6. Answers will vary. However, students should recognize the difference between the role zeros play when calculating products and sums.

Part 2: Scientific Notation and Stinky Feet

1. There are approximately 250,000 sweat glands in our feet. These sweat glands produce a lot of sweat, causing an environment that is rich for bacteria growth. As bacteria grow, they excrete their waste. This waste is what causes the smell. 250,000 sweat glands
2. Answers will vary. However, students should multiply 250,000 by the number of people in your classroom to get the answer.
3. Answers will vary.
4. 318,000,000
5. Multiply the number of people in the United States by 250,000.
6. 2.5×10^5
7. a. $318{,}000{,}000 \times 250{,}000 = 79{,}500{,}000{,}000{,}000$
   b. $(3.18 \times 2.5) \times (10^8 \times 10^5) = 7.95 \times 10^{13}$
8. They are the same.
9. Answers will vary.
10. 1,750,000,000,000,000
11. I could divide the number of feet sweat glands in the world by 250,000.

12. a. $\dfrac{1{,}750{,}000{,}000{,}000{,}000}{250{,}000} = 7{,}000{,}000{,}000$

   b. $\dfrac{1.75}{2.5} \times \dfrac{10^{15}}{10^5} = 0.7 \times 10^{10} = 7.0 \times 10^9$

13. They are the same.
14. Answers will vary.

Day 2: Those Are Some Crazy Numbers!
Part 1: Researching the Facts

Question	Standard Form	Scientific Notation
1. How much does the Earth weigh in tons?	6,585,000,000,000,000,000,000	6.585×10^{21}
2. How much does the United States spend on clothes each day?	$300,000,000	3.0×10^{8}
3. How many words are there in the English language?	1,000,000	1.0×10^{6}
4. How many miles does the Earth travel in five trips around the sun?	2,917,080,000	2.9×10^{9}
5. How many cats are in the United States?	74,000,000 to 96,000,000	7.4×10^{7} to 9.6×10^{7}
6. How many dogs are in the United States?	70,000,000 to 80,000,000	7.0×10^{7} to 8.0×10^{7}
7. How many gallons of saliva does a person produce in their lifetime?	10,000	1.0×10^{4}

Part 2: Applying the Research
1. $(6.585 \times 10^{21})(2.0 \times 10^{3}) = 13.17 \times 10^{24} = 1.317 \times 10^{25}$ pounds
2. $(3.0 \times 10^{8})(3.65 \times 10^{2}) = 10.95 \times 10^{10} = 1.095 \times 10^{11}$ dollars

3. $\dfrac{1.095 \times 10^{11}}{3.18 \times 10^{8}} \approx 0.344 \times 10^{3} \approx 3.44 \times 10^{2}$ dollars

4. $\dfrac{1.0 \times 10^{6}}{6.0 \times 10^{4}} \approx 0.167 \times 10^{2} \approx 1.67 \times 10^{1}$ times more words in the English language than the graduate's vocabulary

5. Answers will vary. However, students should divide 2.9×10^{9} by 5 to determine the distance traveled in 1 year and then multiply that number by their age in years.

6. $(7.4 \times 10^{7}) + (7.0 \times 10^{7}) = 14.4 \times 10^{7}$ or 1.44×10^{8};
$(9.6 \times 10^{7}) + (8.0 \times 10^{7}) = 17.6 \times 10^{7}$ or 1.76×10^{8}
There are 1.44×10^{8} to 1.76×10^{8} cats and dogs in the United States.

7. $(7.4 \times 10^{7}) - (7.0 \times 10^{7}) = 0.4 \times 10^{7}$ or 4.0×10^{6};
$(9.6 \times 10^{7}) - (8.0 \times 10^{7}) = 1.6 \times 10^{7}$
There are 4.0×10^{6} to 1.6×10^{7} more cats than dogs in the United States.

8. $(3.18 \times 10^{8})(1.0 \times 10^{4}) = 3.18 \times 10^{12}$ gallons

Grade 8, Lab #4: Round and Round We Go! (pgs. 79–83)
Day 1: Who Needs Pi?
Part 2: What Does π Really Represent?
1. Answers will vary but should be close to 3.14.
2. They are all approximately 3.14.

3. $C = \pi d;\ \dfrac{C}{d} = \dfrac{\pi d}{d} \longrightarrow \dfrac{C}{d} = \pi$

4. Pi is the ratio of circumference to diameter.
Day 2: Circular Construction
Part 1: Making a Cylinder
6. 5.5 inches
7. 8.5 inches
8. $C = \pi d \longrightarrow 8.5 = 3.14d \longrightarrow 2.71 = d$; The diameter is about 2.71 inches.
Part 2: Making a Cone
3. 5.5 inches
4. 8.5 inches
5. $C = \pi d \longrightarrow 8.5 = 3.14d \longrightarrow 2.71 = d$; The diameter is about 2.71 inches.
Part 3: Making a Sphere
2. $C = \pi d \longrightarrow 8.5 = 3.14d \longrightarrow 2.71 = d$; The diameter is about 2.71 inches.
Day 3: Round and Round We Go!
Part 1: Estimating the Volume of a Cone
2. Answers will vary.
3. a.   I had to fill my cone three times in order to fill the cylinder.
   b.   Answers will vary.
4. They are the same.
5. They are the same.
6. Sample answer: A cone that has the same sized base and height as a cylinder holds $\frac{1}{3}$ the volume that the cylinder holds.
7. $V = \pi r^2 h$
8. I should multiply it by $\frac{1}{3}$, because the cone holds $\frac{1}{3}$ as much as the cylinder. So, the formula for the volume of a cone is $V = \frac{1}{3}\pi r^2 h$.
Part 2: Estimating the Volume of a Sphere
3. Answers will vary.
4. a.   I had to fill my cone one time in order to fill the sphere.
   b.   Answers will vary.
5. They are the same.
6. The radius is about 2.71 ÷ 2 or 1.355 inches.
7. 5.5 ÷ 1.355 ≈ 4; The height of the cone is approximately four times the radius.
8. Because the height is about four times the radius, I can replace the h in the formula with $4r$.
   $V = \frac{1}{3}\pi r^2 h = \frac{1}{3}\pi r^2 (4r) = \frac{4}{3}\pi r^3$
9. Sample answer: A sphere that has the same radius as a cone whose height is about four times its radius holds the same volume that the cone holds.
10. Because the sphere holds the same volume that the cone holds, I can use the formula $V = \frac{4}{3}\pi r^3$ to calculate the volume of a sphere.